T0304580

Reverse Engineering
of Algebraic Inequalities

The second edition of *Reverse Engineering of Algebraic Inequalities* is a comprehensively updated new edition demonstrating the exploration of new physical realities in various unrelated domains of human activity through reverse engineering of algebraic inequalities.

This book introduces a groundbreaking method for generating new knowledge in science and technology that relies on reverse engineering of algebraic inequalities. By using this knowledge, the purpose is to optimize systems and processes in diverse fields such as mechanical engineering, structural engineering, physics, electrical engineering, reliability engineering, risk management and economics. This book will provide the reader with methods to enhance the reliability of systems in total absence of knowledge about the reliabilities of the components building the systems; to develop light-weight structures with very big materials savings; to develop structures with very big load-bearing capacity; to enhance process performance and decision-making; to obtain new useful physical properties; and to correct serious flaws in the current practice for predicting system reliability.

This book will greatly benefit professionals and mathematical modelling researchers working on optimising processes and systems in diverse disciplines. It will also benefit undergraduate students introduced to mathematical modelling, post-graduate students and post-doctoral researchers working in the area of mathematical modelling, mechanical engineering, reliability engineering, structural engineering, risk management, and engineering design.

Reverse Engineering
of Algebraic Inequalities

Exploring New Physical Realities
and Optimizing Products and Processes

Second Edition

Michael T. Todinov

CRC Press
Taylor & Francis Group
Boca Raton London New York

CRC Press is an imprint of the
Taylor & Francis Group, an **informa** business

Designed cover image: Michael T. Todinov

Second edition published 2025
by CRC Press
2385 NW Executive Center Drive, Suite 320, Boca Raton FL 33431

and by CRC Press
4 Park Square, Milton Park, Abingdon, Oxon, OX14 4RN

CRC Press is an imprint of Taylor & Francis Group, LLC

© 2025 Michael T. Todinov

First edition published by CRC Press 2021.

ISBN: 978-1-032-84856-3 (hbk)
ISBN: 978-1-032-85343-7 (pbk)
ISBN: 978-1-003-51776-4 (ebk)

DOI: 10.1201/9781003517764

Typeset in Times
by SPi Technologies India Pvt Ltd (Straive)

Dedication

To Prolet and Marin

Contents

Contents

Preface

An impressive advantage of using algebraic inequalities to model physical systems and processes is that algebraic inequalities do not require specific knowledge of the values or distributions of the variables involved. This advantage permits ranking of systems and processes even when the performance characteristics of their components are unknown.

In a previous publication, the author explored the applications of algebraic inequalities for risk and uncertainty reduction, focusing on a forward approach to generating new knowledge. According to the forward approach, the performance of competing physical systems or processes is compared by deriving and proving relevant algebraic inequalities. However, the potential of this approach is somewhat limited, as it is confined to specific performance characteristics or systems.

This book introduces a reverse engineering approach, which significantly enhances the capabilities of the forward approach. Instead of comparing systems and processes by deriving and proving algebraic inequalities and demonstrating the superiority of one of the design alternatives, the reverse engineering approach works in the opposite direction. It starts with an established algebraic inequality and progresses towards deriving new knowledge about a real system or process. This knowledge is then used to enhance the performance of the system or process.

The reverse engineering of algebraic inequalities is based on the principle of consistency (non-contradiction): *if the variables and the different terms of a correct algebraic inequality can be physically interpreted as parts of a system or process, in the physical world, the system or process exhibit properties or behaviours that are consistent with the prediction of the algebraic inequality.*

The principle of consistency reflects the interconnected nature of reality and all existence. Patterns observed by reverse engineering of a correct algebraic inequality are mirrored and reflected by the physical reality thereby maintaining consistency between the different levels of reality. As a result, reverse engineering of algebraic inequalities generates deep insights into various physical phenomena and processes. These insights open opportunities for enhancing the performance of systems and processes through the discovery of new fundamental properties. Reverse engineering of algebraic inequalities offers a groundbreaking approach to generating new knowledge in science and technology.

Furthermore, depending on the specific interpretation, knowledge applicable to systems and processes from diverse domains can be released from reverse engineering of the same inequality. An important step of the reverse engineering approach is creating, relevant physical meaning for the variables entering the algebraic inequality, followed by a meaningful physical interpretation of the different parts of the inequality and releasing new knowledge.

This book is deeply rooted in the practical applications of algebraic inequalities, particularly in engineering design. Design inequalities are widely used to express constraints that ensure reliable performance of all required functions.

However, as this book demonstrates, the potential applications of algebraic inequalities in engineering extend far beyond defining design constraints. By physically interpreting algebraic inequalities, engineers can derive new knowledge and optimize systems and processes, adding a new dimension to the understanding and application of algebraic inequalities. The practical uses of algebraic inequalities are explored across various fields, including reliability engineering, risk management, mechanical engineering, electrical engineering, civil engineering, economics, operational research, and project management.

Thus, Chapter 4 demonstrates the use of reverse engineering of sub-additive and super-additive algebraic inequalities to derive a powerful method for developing lightweight designs and enhancing the load-bearing capacity of common structures. The essence of the proposed method is consolidating multiple elements loaded in bending into a reduced number of elements with larger and geometrically similar cross sections but a smaller total volume of material. This procedure yields a huge reduction in material usage. For instance, when aggregating ten load-carrying beams into two beams supporting the same total load, the material reduction is more than 1.71 times! In addition, it is demonstrated that consolidating multiple elements loaded in bending into a reduced number of elements with larger, geometrically similar cross sections and with the same total volume of material, leads to a big increase in the load-bearing capacity of the structure. For instance, when aggregating eight cantilevered or simply supported beams into two beams with the same volume of material, the load-bearing capacity until a specified tensile stress increases twice. At the same time, the load-bearing capacity until a specified deflection increases four times.

Despite the widespread use of multiple identical load-carrying elements in engineering and construction, and decades of research, the results from the reverse engineering of algebraic inequalities presented in this book have remained invisible to experts in mechanical engineering, stress analysis, structural engineering, and structural reliability.

However, the reverse engineering of algebraic inequalities extends far beyond this single area of application. Chapters 3 and 10 for example, also explore the reverse engineering of sub-additive and super-additive inequalities. The results are valuable insights for optimizing systems and processes across various fields of science and technology, provided the variables and terms in the inequalities are additive quantities.

Chapters 3, and 10 focus on reverse engineering of sub-additive and super-additive inequalities to maximize: total power output from a voltage source, energy stored in a capacitor, elastic strain energy stored during tension and bending, performance of a filter, profit from investment and the mass of deposited substance during electrolysis. Reverse engineering of sub-additive and super-additive inequalities was also used to minimize: the drag force during motion in a viscous fluid, the formation of undesirable brittle phases during phase transformations and the accumulated kinetic energy during inelastic impact.

Chapter 5 demonstrates the use of reverse engineering of algebraic inequalities to improve the reliability of series-parallel systems without any knowledge of components' reliabilities. Specifically, this chapter establishes that for series-parallel systems, an asymmetric arrangement of interchangeable redundancies consistently results in higher system reliability compared to a symmetric arrangement, regardless of the individual reliability values of the components. The proposed technique based on asymmetric arrangement of redundancies is not associated with any implementation costs and can be used for improving reliability and reducing the risk of failure in various unrelated domains.

Through a reverse engineering of an algebraic inequality, Chapter 7 reveals that a prediction of system reliability on demand based on average reliabilities on demand of components *is a fundamentally flawed approach in system reliability theory*. Despite decades of system reliability research, these results, derived from reverse engineering of algebraic inequalities, remained unnoticed by system reliability experts.

The reverse engineered algebraic inequalities demonstrated that assuming average component reliabilities on demand entails an overestimation of the system reliability on demand for systems with components logically arranged in series and series-parallel and underestimation of the reliability on demand for systems with components logically arranged in parallel.

Chapter 8 explores the reverse engineering of the inequality of additive ratios and presents an important result in probability theory. The inequality of additive ratios can be used to increase the probability of an event occurring within a set of mutually exclusive and exhaustive events. Despite research spanning many decades, this key result, derived from reverse engineering of algebraic inequalities, remained elusive to probability experts.

The chapter also explores applications related to increasing the mass of substance deposited during electrolysis, optimal resource allocation, and presents a key result related to the deflection of elastic elements connected in series and parallel.

The reverse engineering of various algebraic inequalities in Chapter 9 has led to a highly counter-intuitive finding regarding the procurement of components from suppliers with unknown proportions of reliable components. Despite having no knowledge about the proportions of reliable components from each supplier, purchasing all components from the same supplier results in the highest probability that all components will be reliable.

Chapter 11 covers reverse engineering of algebraic inequalities to enhance decision-making processes. By analysing simple algebraic inequalities, it draws important conclusions related to predicting the ranking of sequential random events. The chapter also covers algebraic inequalities whose reverse engineering yields tight bounds on the probability of selecting components of the same or different varieties.

Chapter 12 demonstrates that algebraic inequalities can also be reverse engineered in terms of potential energy within a system. In such cases, the constant on the right-hand side of the inequality represents the system's state of stable equilibrium, corresponding to minimal potential energy. From the equilibrium conditions associated with this stable state, several valuable relationships can be deduced, allowing determination of the unknown lower bound without resorting to complex models.

This book is a pioneering publication on generating new knowledge through reverse engineering of algebraic inequalities. For decades, many of the results presented in this book eluded domain experts, demonstrating that they could not be found without employing reverse engineering of algebraic inequalities. By using the principle of consistency, the book illustrates the profound connection between physical reality and mathematics. The findings presented support the view that real physical phenomena and processes inherently follow paths that align with correct abstract algebraic inequalities. This alignment ensures that no contradictions arise with algebraic inequalities, whose variables and terms correspond to the controlling factors driving these phenomena and processes.

The book can be used as a primary source for courses on algebraic inequalities and their applications, and as essential reading for researchers aiming to make a significant impact by interpreting algebraic inequalities in the context of physical systems and processes. In addition, the book offers excellent resources for applications in courses and projects related to Mathematical Modelling, Applied Mathematics, Reliability Engineering, Mechanical Engineering, and Electrical Engineering.

In conclusion, I would like to express my gratitude to Katya Porter, the mechanical engineering editor at CRC Taylor and Francis Group, Shatakshi Singh, the editorial assistant for mechanical engineering, and Thivya Vasudevan, project Manager at Straive (SPi Global) for their invaluable help and cooperation. I also extend my thanks to my academic colleagues for their insightful comments on various aspects of the presented results.

Lastly, I acknowledge the immense support and encouragement from my wife, Prolet, during the preparation of this book.

Michael T. Todinov
Oxford, June, 2024

Author's Biography

Michael T. Todinov is a professor of mechanical engineering at Oxford Brookes University, UK, where he teaches reliability engineering, engineering mathematics, and advanced stress analysis. He holds a PhD in mechanical engineering and a higher doctorate, equivalent to a DSc, in mathematical modeling.

Prof. Todinov has pioneered innovative research in several areas, including reverse engineering of algebraic inequalities, domain-independent methods for reliability improvement, analysis and optimization of repairable flow networks, reliability analysis based on the cost of failure, and probabilistic fracture controlled by defects.

Michael Todinov is a recipient of a prestigious award from the Institution of Mechanical Engineers (UK) for his work on risk reduction in mechanical engineering.

1 Fundamental Approaches in Modelling Real Systems and Processes by Using Algebraic Inequalities

The Principle of Consistency for Algebraic Inequalities

1.1 ALGEBRAIC INEQUALITIES AND THEIR GENERAL APPLICATIONS

Algbraic inequalities have been used extensively in mathematics and a number of useful non-trivial algebraic inequalities and their properties have been well documented (Bechenbach and Bellman, 1961; Cloud et al., 1998; Engel, 1998; Hardy et al., 1999; Pachpatte, 2005; Steele, 2004; Kazarinoff, 1961; Sedrakyan & Sedrakyan, 2010; Yong and Bin, 2016). A comprehensive overview on the use of inequalities in mathematics has been presented in (Fink, 2000).

For a long time, algebraic inequalities have been used in mathematics to express error bounds in approximations and constraints in linear programming models. In reliability and risk research, inequalities have been used as a tool for characterisation of reliability functions (Ebeling, 1997; Xie and Lai, 1998; Makri and Psillakis, 1996; Hill et al., 2013; Berg and Kesten, 1985; Kundu and Ghosh, 2017; Dohmen, 2006) and for reducing uncertainty and risk (Todinov, 2020a). Applications of inequalities have been considered in physics (Rastegin, 2012) and engineering (Cloud et al., 1998; Samuel and Weir, 1999).

In engineering design, design inequalities have been used widely to express design constraints guaranteeing that the design will perform its required function (Samuel and Weir, 1999). This book shows that the application potential of algebraic inequalities in engineering is far reaching and certainly not restricted to specifying design constraints and error bounds.

DOI: 10.1201/9781003517764-1

The method of algebraic inequalities is a domain-independent method which derives from the domain-independent nature of mathematics. Applications of algebraic inequalities can be demonstrated in such diverse domains as: mechanical engineering, electrical engineering, optimisation, operational research, project management, economics, decision-making under uncertainty, manufacturing and quality control.

1.2 ALGEBRAIC INEQUALITIES AS A DOMAIN-INDEPENDENT METHOD FOR REDUCING UNCERTAINTY AND OPTIMISING THE PERFORMANCE OF SYSTEMS AND PROCESSES

The conventional approaches for handling uncertainty in reliability engineering, for example, rely heavily on probabilities (Ramakumar, 1993; Lewis, 1996; Ebeling, 1997; Hoyland and Rausand, 1994; Dhillon, 2017; Todinov 2002a, 2006a,b). These approaches effectively deal with structured uncertainty, and the required probabilities are usually assessed by using a data-driven method or a Bayesian, subjective probability method.

A major deficiency of the data-driven approach is that probabilities cannot always be meaningfully defined. To be capable of making predictions, the data-driven approach needs past failure rates. Models based on failure rate data collected for a particular environment (temperature, humidity, pressure, vibrations, corrosive agents, etc.) however, give poor predictions for the time to failure in a different environment. Furthermore, the failure rates are average quantities, and as it has been demonstrated in (Todinov, 2023c), the average failure rates yield poor predictions of the system's reliability.

The deficiencies of the data-driven approach in reliability engineering, cannot be rectified by using the Bayesian approach which is not so critically dependent on the availability of past failure data since it uses subjective, knowledge-based probabilities expressing the degree of belief about the outcome of a random event (Winkler, 1996). The subjective probabilities are subsequently updated as new experimental evidence becomes available (Ang and Tang, 2007). The Bayesian approach however, depends on a selected probability model that may not be relevant to the modelled phenomenon/process (Aven, 2017). In addition, the assigned subjective probabilities depend on the available knowledge and vary significantly among the assessors. In this respect, weak background knowledge underlying the assigned subjective probabilities often results in poor predictions. Although the info-gap theory (Ben-Haim, 2005) deals with unstructured uncertainty by not making probability distribution assumptions, it still requires assumptions to be made about designer's best estimate.

In using algebraic inequalities to handle unstructured uncertainty, there is no need to assign frequentist probabilities, subjective probabilities or any particular probabilistic models. Algebraic inequalities *do not require knowledge related to the distributions of the variables entering the inequalities or any other assumptions,* and this makes the method of algebraic inequalities ideal for handling deep uncertainty associated with components, properties and values of control parameters.

Although the probabilities remain unknown, the algebraic inequalities can still establish the intrinsic superiority of one of the competing options. In this respect, algebraic inequalities avoid a major difficulty in the conventional models for handling uncertainty – lack of meaningful specification of frequentist probabilities or weak knowledge behind the assigned subjective probabilities and probabilistic models. Furthermore, the method of algebraic inequalities is capable of discovering new properties related to systems and processes by physical interpretation of the inequalities.

While reliability and risk assessment (Henley and Kumamoto, 1981; Kaplan and Garrick, 1981; Vose, 2000; Aven, 2003; Todinov, 2007) are truly domain-independent areas, this cannot be stated about the equally important areas of reliability improvement and risk reduction. For decades, the reliability and risk science failed to appreciate and emphasize that reliability improvement, risk and uncertainty reduction are underpinned by general principles that work in many unrelated domains.

As a consequence, *methods for measuring and assessing reliability, risk, and uncertainty were developed, but not domain-independent methods for improving reliability and reducing risk and uncertainty, which could provide direct input to the design process.* Indeed, in standard textbooks on mechanical engineering and design of machine components (French, 1999; Samuel and Weir, 1999; Mott et al., 2018; Norton, 2006; Pahl et al., 2007; Gullo and Dixon, 2018; Thompson, 1999; Childs, 2014; Budynas and Nisbett, 2015; Maier et al. 2022; Ugural, 2022) for example, there is practically no discussion of generic (domain-independent) methods for reliability improvement and risk and uncertainty reduction.

However, the TRIZ problem-solving framework (Altshuller, 1999; Orloff, 2006; Rantanen and Domb, 2008; Koziołek et al., 2018), widely adopted by companies and researchers around the world, clearly demonstrated the advantages of using generic principles in resolving technical contradictions and driving innovation. Consequently, domain-independent methods for improving reliability promote rapid mental mapping and bolstering intuition which leads to surprising breakthroughs and swift solutions for challenging problems.

The problem is that the current approach to reliability improvement and risk reduction almost solely relies on knowledge from a specific domain and is conducted exclusively by experts in that domain. This creates the incorrect perception that effective reliability improvement and risk reduction can be delivered only by using methods offered by the specific domain, without resorting to general risk reduction methods and principles.

This incorrect perception resulted in ineffective reliability improvement and risk reduction, the loss of valuable opportunities for reducing risk and repeated 'reinvention of the wheel'. Current technology changes so fast that the domain-specific knowledge related to reliability improvement and risk reduction is often outdated almost as soon as it is generated. In contrast, the domain-independent methods for reliability improvement are higher-order generic methods that permit universal application in new, constantly changing situations and circumstances.

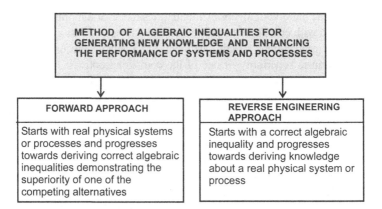

FIGURE 1.1   Two basic approaches of the method of algebraic inequalities for generating new knowledge and enhancing system and process performance.

A central theme in the domain-independent approach for reliability improvement and risk reduction introduced in (Todinov, 2019a,b) is the concept that risk reduction is underlined by common domain-independent principles which, if combined with knowledge from the specific domain, (i) provide key inputs to the design process; (ii) avoid loss of opportunities for improving reliability and reducing risk and (iii) provide effective risk and uncertainty reduction across unrelated domains of human activity. As part of the domain-independent approach for reliability improvement and risk reduction, a number of new domain-independent methods have been introduced: *separation, segmentation, self-reinforcement, introducing deliberate weaknesses, inversion, reducing the rate of damage accumulation, increasing the level of balancing, asymmetric arrangement of interchangeable redundancies and substitution*. The method of algebraic inequalities (Todinov, 2020a, 2020b) is another powerful domain-independent method for improving the reliability of systems and processes. There are two major approaches of the method of algebraic inequalities: (i) *forward approach*, which starts with real systems and processes and progresses towards deriving correct algebraic inequalities demonstrating the superiority of one of the competing alternatives and (ii) *inverse (reverse engineering) approach*, which consists of interpreting a correct abstract inequality and inferring from it unknown properties related to a real physical system or process (Todinov, 2021a) (Figure 1.1).

1.3 FORWARD APPROACH TO MODELLING AND OPTIMISATION OF REAL SYSTEMS AND PROCESSES BY USING ALGEBRAIC INEQUALITIES

The forward approach includes several basic steps (Figure 1.2): (i) analysis of the physical system or process and the available alternatives; (ii) conjecturing inequalities ranking the competing alternatives; (iii) testing and proving rigorously the

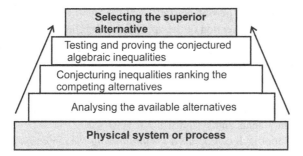

FIGURE 1.2 Forward approach to modelling real systems and processes by using algebraic inequalities.

conjectured inequalities and (iv) selecting the superior competing alternative. The forward approach for generating knowledge and improving systems and process performance has already been explored in (Todinov, 2019c, d; 2020a, 2020b, 2020c, 2020d). By following this approach, algebraic inequalities for example, can be used to *rank systems with unknown reliabilities of their components.*

The generic strategy in comparing the reliabilities of competing systems starts with building the functional diagrams of the physical systems, creating reliability networks based on the functional diagrams, deriving expressions for the system reliability of the competing alternatives, conjecturing inequalities related to the competing alternatives, testing the conjectured inequalities by using Monte Carlo simulation and developing a rigorous proof by using some combination of analytical techniques for proving algebraic inequalities (Figure 1.3).

As a result, the reliabilities of two systems can be compared in the absence of knowledge about the reliabilities of the separate components or in the presence of partial knowledge only. Partial knowledge is present if, for example, it is known that a particular component is older (less reliable) than another component of the same type.

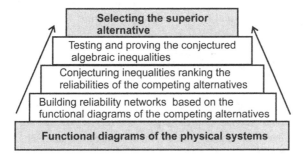

FIGURE 1.3 Forward approach used for ranking the reliabilities of systems with unknown reliabilities of their components.

Algebraic inequalities permit not only ranking two systems in terms of reliability but also the identification of the optimal (most reliable) system if all available topologies are available. Such an application of inequalities to achieve system reliability optimisation has been discussed in Todinov (2020b).

The forward approach can also be used *for determining upper and lower bounds for critical properties of systems and processes*. In this way, inequalities can be used for reducing the worst-case variation of properties and improving the reliability of components and systems.

The forward approach can be used *for minimising the deviation of reliability-critical parameters from their required values* and *for maximising the system reliability*. These applications have been demonstrated in (Todinov, 2020a, 2020b).

1.4 INVERSE (REVERSE ENGINEERING) APPROACH TO MODELLING AND GENERATING NEW KNOWLEDGE BY PHYSICAL INTERPRETATION OF ALGEBRAIC INEQUALITIES

The treatment related to the applications of algebraic inequalities (Todinov, 2020a) for reducing uncertainty and risk was based exclusively on the forward approach, outlined in Section 1.3. The forward approach is powerful but its potential for generating new knowledge is somewhat limited. The comparison of competing systems/processes does not always yield algebraic inequalities that hold for all values of the controlling variables. In some cases, the conjectured inequality holds for some set of values of the controlling variables, but for another set of values, it does not hold. As a result, in some cases, no intrinsically more reliable system or process emerges as a result of applying the forward approach. Furthermore, the potential for generating new knowledge of the forward approach is confined to the specific systems and processes that are compared.

These limitations are overcome by the inverse (reverse engineering) approach, which has been proposed in (Todinov, 2021a). The reverse engineering approach starts with a correct algebraic inequality, progresses through attaching relevant meaning physical for the variables entering the inequality, followed by a meaningful physical interpretation of the different parts of the inequality which links it to a real systemor process, and ends with formulating undiscovered properties/ knowledge about the physical system or process (Figure 1.4).

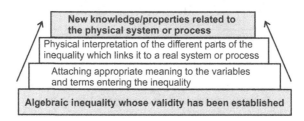

FIGURE 1.4 Reverse engineering approach to deriving new knowledge by using algebraic inequalities.

The reverse engineering approach, which is at the focus of this book, is founded on the observation that abstract algebraic inequalities contain useful quantitative knowledge that can be released through their meaningful physical interpretation. Furthermore, depending on the specific interpretation, knowledge applicable to systems and processes from diverse domains can be released from the same inequality. In addition, the reverse engineering approach does not require or imply any forward analysis of existing systems or processes.

The reverse engineering of algebraic inequalities reflects the interconnected nature of reality and all existence. Patterns observed by reverse engineering of a correct algebraic inequality are mirrored and reflected by the physical reality thereby maintaining consistency between the different levels of reality. As a result, the reverse engineering of algebraic inequalities generates insights into various physical phenomena and processes. which opens opportunities for enhancing the performance of systems and processes.

The reverse engineering of algebraic inequalities is a ground-breaking approach to generating new knowledge in science and technology and discovering new fundamental properties of physical systems.

The initial step of the inverse approach is testing a conjectured algebraic inequality (e.g. by using Monte Carlo simulation) (Figure 1.4). If no combinations of input data contradicting the inequality have been found during the testing stage, the conjectured inequality is plausible and a rigorous proof is attempted. Next, relevant physical meaning for the variables entering the inequality is created, followed by a physical interpretation of its parts (Figure 1.4).

A trivial algebraic inequality can certainly be physically interpreted, but the knowledge extracted from this interpretation can also be reached intuitively. In contrast, the interpretation of a non-trivial algebraic inequality, for example, the inequality $x^2 + y^2 + z^2 \geq xy + yz + zx$, where x, y and z are any real numbers, leads to deep insights that are not at all obvious and cannot be reached intuitively. As demonstrated in this book, interpreting non-trivial inequalities often yields counter-intuitive insights.

The key idea of this book is that non-trivial algebraic inequalities can be physically interpreted and the new knowledge extracted from the physical interpretation can be used for optimising systems and processes in diverse areas of science and technology. The knowledge extracted from the interpretation of non-trivial inequalities is non-trivial and cannot be reached intuitively.

Interpretation of abstract inequalities helps to find overlooked useful properties in such mature fields like mechanical engineering, electrical engineering, reliability engineering and risk management.

The reverse engineering approach always leads to new results, as long as a meaningful physical interpretation of the variables and the different parts of the inequalities can be done. This is because this approach is rooted in the principle of consistency (non-contradiction): *if the variables and the different terms of a correct algebraic inequality can be interpreted as controlling factors or attributes of a physical system or process, in the physical world, the system or process exhibits properties and behaviour that are consistent with the prediction of the*

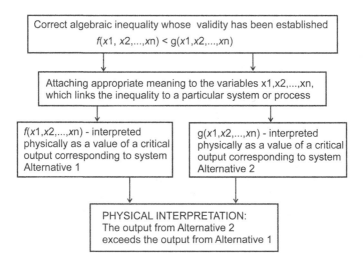

FIGURE 1.5 *Optimising a system or process by reverse engineering of algebraic inequalities.*

algebraic inequality. In short, the realization of the process/experiment yields results that do not contradict the algebraic inequality.

As a result, the reverse engineering of a valid algebraic inequality leads to the projection of a new physical reality which is characterised by a distinct signature – the algebraic inequality itself.

A central idea in optimising designs by using algebraic inequalities is to interpret the left- and right-hand side of a correct algebraic inequality as a particular output related to two different design options, delivering the same required function. The algebraic inequality then establishes the superiority of one of the compared design options with respect to the chosen output (Figure 1.5).

1.5   THE PRINCIPLE OF CONSISTENCY FOR ALGEBRAIC INEQUALITIES

The link between physical reality and algebraic inequalities can be demonstrated with the next simple example. Consider the common abstract inequality

$$a^2 + b^2 \geq 2ab \tag{1.1}$$

which is true for any real numbers a, b because the inequality can be obtained from the obvious inequality $(a - b)^2 \geq 0$. Adding $2ab$ to both sides of inequality (1.1) will not change its direction and the result is the inequality

$$\left(a + b\right)^2 \geq 4ab \tag{1.2}$$

Consider the positive quantities a, b ($a > 0$, $b > 0$). Dividing both sides of inequality (1.2) by the positive value ($a + b$) does not alter the direction of the inequality:

$$a + b \geq 4 \frac{ab}{a + b} \quad (1.3)$$

Dividing the numerator and denominator of the right-hand side of (1.3) by ab gives

$$a + b \geq 4 \frac{1}{1/a + 1/b} \quad (1.4)$$

The left- and right-hand side of inequality (1.4) can be physically interpreted by interpreting the variables a and b as electrical resistances. The left-hand side of inequality (1.4) can be interpreted as the equivalent resistance of two elements connected in series (Figure 1.6a). The right-hand side of inequality (1.4) can be interpreted as the equivalent resistance of the same elements connected in parallel, multiplied by 4 (Figure 1.6b).

Inequality (1.4) predicts that *the equivalent resistance of two elements connected in series is at least four times greater than the equivalent resistance of the same elements connected in parallel, irrespective of the individual resistances of the elements.*

If physical measurements of the equivalent resistances of the arrangements in Figure 1.6a and b are conducted, they will only confirm the prediction from the algebraic inequality: that for any combination of values a and b for the resistances of the two elements, the equivalent resistance in series is always at least four times greater than the equivalent resistance of the same elements in parallel.

The examples in this book demonstrate a similar perfect agreement between predictions from abstract inequalities and the properties and behaviour of physical systems and processes. In rationalizing the agreement of experimental results with predictions from abstract inequalities, a natural question arises: why does the real world align its responses with what an abstract inequality predicts?

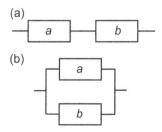

FIGURE 1.6 Parts of electrical circuit demonstrating the two-way connection between algebraic inequalities and physical reality; (a) resistors connected in series and (b) the same resistors connected in parallel.

The debate about the unreasonable effectiveness of mathematics in the natural sciences started with Wigner's famous paper in 1960 (Wigner, 1960). Recent decades have seen a continuation of this discourse (Penrose, 1989; Livio, 2009; Vedral, 2010; Tegmark, 2014), relating to the ancient question dating back to Pythagoras and Plato: whether mathematics is a human invention or expresses truths with independent existence that are discovered. The positions in this vigorous debate range from 'mathematics is a pure human invention,' a set of rules for manipulating symbols, to the view that 'reality is mathematics' (Tegmark, 2014).

The results presented in this book support the view that physical phenomena/ processes take (follow) paths that are consistent with the predictions of correct abstract algebraic inequalities whose variables and different parts correspond to key parameters controlling these phenomena/processes. This is the essence of the principle of consistency which underlies deriving new knowledge from reverse engineering of algebraic inequalities. It seems that mathematics is interwoven in the fabric of the physical world and the course of physical phenomena and processes is inherently consistent with the relevant algebraic inequalities.

The principle of consistency between the predictions of correct algebraic inequalities and the development of real processes not only infers the existence of particular properties but also forbids the existence of properties that are not in agreement with the relevant algebraic inequalities. To illustrate this point, consider two objects and a physical process where the probability of appearance of a particular feature X in a selected object is p. Consequently, the probability of non-appearance of the feature X in the selected object is $1 - p$. In addition, the probability of appearance/non-appearance of the feature in one of the objects does not depend on the appearance/non-appearance of the feature in the other object. This means that the appearance/non-appearance of the feature in the individual objects are statistically independent events.

In this case, the probability that the two objects will either both have the feature X or both will not have it is

$$P\left(X \cap X \cup \bar{X} \cap \bar{X}\right) = p^2 + \left(1-p\right)^2 \tag{1.5}$$

The probability that one of the objects will have the feature and the other object will not is

$$P\left(X \cap \bar{X} \cup \bar{X} \cap X\right) = p\left(1-p\right) + \left(1-p\right)p = 2p\left(1-p\right) \tag{1.6}$$

Suppose that it has been conjectured that the appearance/non-appearance of the feature in the individual objects are statistically independent events. In this case, the inequality

$$p^2 + \left(1-p\right)^2 \geq 2p\left(1-p\right) \tag{1.7}$$

follows directly from the obvious inequality $[p - (1 - p)]^2 \geq 0$, irrespective of the probability p. If the appearance/non-appearance of the feature in each of the objects are statistically independent events, inequality (1.7) must hold.

Let a particular experiment yield empirically that the probability of both objects having the feature or both objects not having the feature is smaller than the probability that one object will have the feature and the other object will not: $P\left(X \cap X \cup \bar{X} \cap \bar{X}\right) < P\left(X \cap \bar{X} \cup \bar{X} \cap X\right)$. In this case, inequality (1.7) does not hold for that experiment, which means that the appearance and non-appearance of the feature in the individual objects cannot be statistically independent events.

1.6 KEY STEPS IN THE PHYSICAL INTERPRETATION OF ALGEBRAIC INEQUALITIES

The key steps of the physical interpretation of algebraic inequalities are: (i) identifying classes of abstract inequalities which permit interpretation that can be linked easily with a real system or process and (ii) identifying transformations that make the inequality physically interpretable.

Very important candidates for reverse engineering are the wide class of inequalities based on *super-additive* and *sub-additive* functions. If a particular sub-additive or super-additive function measures the effect of a specific additive controlling factor, the sub-additive inequality, for example, suggests the possibility of significantly increasing the effect of the controlling factor by segmenting it.

Additive quantities can be found in all areas of science and engineering. Additive quantities change with changing the size of their supporting objects/systems. Examples of additive quantities are mass, weight, amount of substance, number of particles, volume, distance, energy (kinetic energy, gravitational energy, electric energy, elastic energy, surface energy, internal energy), work, power, heat, force, momentum, electric charge, electric current, heat capacity, electric capacity, resistance (when the elements are in series), enthalpy and fluid flow.

The reverse engineering of algebraic inequalities however, is not restricted to inequalities that only provide segmentation or segregation of additive controlling factors. For example, the reverse engineering of the well-known arithmetic mean – harmonic mean algebraic inequality in Chapter 3 helped reveal important properties of parallel and series arrangements of mechanical, electrical and thermal components. In addition, the reverse engineering of the well-known arithmetic mean–geometric mean algebraic inequality in Chapter 7 helped reveal fundamental flaws in the widely adopted standard technique for predicting system reliability. The reverse engineering of a new class of algebraic inequalities in Chapter 5 helped significantly improve the intrinsic reliability of series-parallel systems by asymmetric arrangement of the interchangeable redundancies.

Knowledge released from reverse engineering of algebraic inequalities is relevant to various domains and helped find overlooked properties and opportunities for enhancing systems performance in mature fields such as reliability engineering, risk management, mechanical engineering, structural engineering, electrical engineering and operations research.

The potential applications of algebraic inequalities are very broad, but the areas of risk and reliability are particularly important application domains. Thus, new knowledge generated in the area of reliability and risk through interpretation of algebraic inequalities provides the opportunity to construct systems with superior intrinsic reliability in the absence of knowledge related to the failure rates of the components building the system.

In the area of mechanical and structural engineering, generating new knowledge through reverse engineering of sub-additive and super-additive algebraic inequalities, as demonstrated in Chapter 4, provided the opportunity to develop lightweight structures with huge material savings and tremendously enhanced load-carrying capacity. In addition, as demonstrated in Chapter 3, the reverse engineering of algebraic inequalities helped maximise the capacity for elastic strain energy accumulation of common mechanical assemblies.

In the field of electrical engineering, new knowledge generated through the reverse engineering of algebraic inequalities has provided the possibility to maximize the power output in electric circuits and the energy stored in capacitors.

In the area of risk management, knowledge generated through reverse engineering of algebraic inequalities provided the opportunity to maximise the profits from an investment, avoid risk underestimation and maximise the likelihood of correct ranking the magnitudes of sequential uncertain events.

The new results derived by reverse engineering of algebraic inequalities have not been stated in any publication in the mature fields of mechanical engineering, reliability engineering, structural engineering, risk management and electrical engineering which demonstrates that the lack of knowledge of the method of algebraic inequalities made these important results invisible to domain experts.

2 Basic Algebraic Inequalities Used in Reverse Engineering and Their Properties

2.1 BASIC ALGEBRAIC INEQUALITIES USED FOR PROVING OTHER INEQUALITIES

2.1.1 FUNDAMENTAL PROPERTIES AND PROOF TECHNIQUES FOR ALGEBRAIC INEQUALITIES

Inequalities are statements about expressions or numbers which involve the symbols '$<$' (less than), '$>$' (greater than), '$\leq$' (less than or equal to) or '$\geq$' (greater than or equal to). The basic rules related to handling algebraic inequalities can be summarised as follows:

a. For any real numbers a and b, exactly one of the following holds:

$$a < b, a = b, a > b.$$

b. If $a > b$ and $b > c$ then $a > c$;

c. If $a > b$, adding the same number c to both sides of the inequality does not alter its direction: $a + c > b + c$;

d. Multiplying both sides of an inequality by (-1) reverses the direction of the inequality:

if $a > b$ then $-a < -b$; if $a < b$ then $-a > -b$;

e. If $a > 0$ and $b > 0$ then $ab > 0$.

By using the basic rules, the next basic properties can be established (see Todinov 2020a for more details and proofs):

i. For any real number x, $x^2 \geq 0$. The equality holds if and only if $x = 0$.

ii. If $x > y$ and $t > 0$, then $xt > yt$ and $x/t > y/t$. From this property, it follows that if $0 < x < 1$ then $x^2 < x$.

iii. If $x > y > 0$ and $u > v > 0$, then $xu > yv$ and $x/v > y/u$. From this property, it follows that if $a > b > 0$ then $a^2 > b^2$.

iv. If $x > 0$, $y > 0$, $x \neq y$ and $x^2 > y^2$, then $x > y$.

DOI: 10.1201/9781003517764-2

v. If a strictly increasing function is applied to both sides of an inequality, the inequality will still hold. Applying a strictly decreasing function to both sides of an inequality reverses the direction of the inequality. Thus, from $x > y > 0$, it follows that $\ln x > \ln y$ and $x^n > y^n$ where $n > 0$. From $x > y > 0$, it follows that, $x^{-n} < y^{-n}$ where $n > 0$.

vi. If $a > b > 0$ and p and q are positive real numbers, then $a^{p/q} > b^{p/q}$.

A comprehensive treatment of various techniques of proving algebraic inequalities has been presented in (Todinov, 2020a) which features the following techniques for proving algebraic inequalities:

- Direct algebraic manipulation and analysis;
- Using mathematical induction;
- Using the properties of convex/concave functions;
- Segmentation through basic inequalities;
- Transforming algebraic inequalities to known basic inequalities;
- Strengthening the inequalities;
- Exploiting the symmetry of the variables entering the inequality;
- Using the properties of sub-additive and super-additive functions;
- Using substitution;
- Exploiting homogeneity;
- Using derivatives.

Most of these techniques are used for proving the inequalities presented in this book.

2.1.2 CAUCHY-SCHWARZ INEQUALITY

An important basic algebraic inequality with numerous applications is the Cauchy-Schwarz inequality (Cauchy, 1821; Steele, 2004) which states that for the sequences of real numbers $a_1, a_2, \ldots, a_n$ and $b_1, b_2, \ldots, b_n$, the following inequality holds:

$$\left(a_1 b_1 + a_2 b_2 + \ldots + a_n b_n\right)^2 \le \left(a_1^2 + a_2^2 + \ldots + a_n^2\right)\left(b_1^2 + b_2^2 + \ldots + b_n^2\right) \quad (2.1)$$

Equality holds if and only if, for any $i \ne j$, $a_i b_j = a_j b_i$ are fulfilled.

The Cauchy-Schwarz inequality is a very powerful inequality and many algebraic inequalities can be proved by reducing them to the Cauchy-Schwarz inequality through appropriate substitutions. The Cauchy-Schwarz inequality (2.1) can be proved by using direct algebraic manipulation and analysis based on the properties of the quadratic trinomial. Consider the expression

$$y = \left(a_1 t + b_1\right)^2 + \left(a_2 t + b_2\right)^2 + \ldots + \left(a_n t + b_n\right)^2 \quad (2.2)$$

For any sequences of real numbers $a_1, a_2, \ldots, a_n$ and $b_1, b_2, \ldots, b_n$, y is non-negative ($y \geq 0$) for any chosen value of t. Expanding the right-hand side of (2.2) and collecting the coefficients in front of t^2 and t, gives the quadratic trinomial

$$y = \left(a_1^2 + a_2^2 + \ldots + a_n^2\right)t^2 + 2\left(a_1b_1 + a_2b_2 + \ldots + a_nb_n\right)t + \ldots + b_1^2 + b_2^2 + \ldots + b_n^2 \quad (2.3)$$

with respect to t. In Equation (2.3), y is non-negative ($y \geq 0$) for any chosen value of t only if $D \leq 0$, where

$$D = \left(a_1b_1 + a_2b_2 + \ldots + a_nb_n\right)^2 - \left(a_1^2 + a_2^2 + \ldots + a_n^2\right)\left(b_1^2 + b_2^2 + \ldots + b_n^2\right)$$

is the discriminant of the quadratic trinomial. Therefore, the condition

$$\left(a_1b_1 + a_2b_2 + \ldots + a_nb_n\right)^2 - \left(a_1^2 + a_2^2 + \ldots + a_n^2\right)\left(b_1^2 + b_2^2 + \ldots + b_n^2\right) \leq 0 \quad (2.4)$$

must hold for a non-negative y. Condition (2.4), however, is identical to the Cauchy-Schwarz inequality (2.1) which completes the proof.

2.1.3 CONVEX AND CONCAVE FUNCTIONS. JENSEN INEQUALITY

A function $f(x)$ with a domain $[a, b]$ is said to be convex (Figure 2.1a), if for all values x_1 and x_2 in its domain ($x_1, x_2 \in [a, b]$), the next inequality holds:

$$f\left(wx_1 + (1-w)x_2\right) \leq wf(x_1) + (1-w)f(x_2) \quad (2.5)$$

where $0 \leq w \leq 1$.

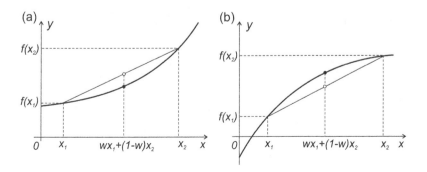

FIGURE 2.1   (a) A convex and (b) a concave function of a single argument.

If $f(x)$ is twice differentiable for all $x \in (a, b)$, and if the second derivative is non-negative ($f''(x) \geq 0$), the function $f(x)$ is convex on (a, b). For any convex function $f(x)$, the Jensen's inequality (Steele, 2004) states that

$$f\left(w_1 x_1 + w_2 x_2 + \ldots + w_n x_n\right) \leq w_1 f\left(x_1\right) + w_2 f\left(x_2\right) + \ldots + w_n f\left(x_n\right) \qquad (2.6)$$

where w_i ($i = 1, \ldots, n$) are numbers (weights) that satisfy $0 \leq w_i \leq 1$ and $w_1 + w_2 + \ldots + w_n = 1$.

If the function $f(x)$ is strictly convex (second derivative is positive), equality in (2.6) is attained only for $x_1 = x_2 = \ldots = x_n$.

A function $f(x)$ with a domain $[a,b]$ is said to be concave (Figure 2.1b), if for all values x_1 and x_2 in its domain ($x_1, x_2 \in [a, b]$), the next inequality holds:

$$f\left(w x_1 + \left(1 - w\right) x_2\right) \geq w f\left(x_1\right) + \left(1 - w\right) f\left(x_2\right) \qquad (2.7)$$

where $0 \leq w \leq 1$.

For any concave function $f(x)$, the Jensen's inequality states that

$$f\left(w_1 x_1 + w_2 x_2 + \ldots + w_n x_n\right) \geq w_1 f\left(x_1\right) + w_2 f\left(x_2\right) + \ldots + w_n f\left(x_n\right) \qquad (2.8)$$

where w_i ($i = 1, \ldots, n$) are numbers (weights) that satisfy $0 \leq w_i \leq 1$ and $w_1 + w_2 + \ldots + w_n = 1$.

If $f(x)$ is twice differentiable for all $x \in (a, b)$, and if the second derivative is not positive ($f''(x) \leq 0$), the function $f(x)$ is concave on (a, b). If the function $f(x)$ is strictly concave (second derivative is negative), equality in (2.8) is attained only for $x_1 = x_2 = \ldots = x_n$.

Proving an inequality by using the properties of concave functions will be demonstrated by proving the inequality:

$$\left(\frac{m_1 x_1 + m_2 x_2 + \ldots + m_n x_n}{M}\right)^M \geq x_1^{m_1} x_2^{m_2} \ldots x_n^{m_n} \qquad (2.9)$$

where $m_1, \ldots, m_n$ and $x_1, \ldots, x_n$ are positive values, and $M = \sum\limits_{i=1}^{n} m_i$.

Consider the function $\ln x$ which is concave for positive values $x > 0$ because its second derivative $(\ln x)'' = -1/x^2$ is negative. According to the Jensen's inequality, for concave functions,

$$\ln\left(w_1 x_1 + w_2 x_2 + \ldots + w_n x_n\right) \geq w_1 \ln x_1 + w_2 \ln x_2 + \ldots + w_n \ln x_n$$

is fulfilled, where $w_i = m_i/M$. Note that w_i are treated as weights because $0 < w_i < 1$ and $\sum\limits_{i=1}^{n} w_i = 1$.

Using the properties of the logarithms

$$\ln\left(w_1 x_1 + w_2 x_2 + \ldots + w_n x_n\right) \geq \ln\left(x_1^{w_1} x_2^{w_2} \ldots x_n^{w_n}\right)$$

Because both sides of the last inequality are positive, and e^x is an increasing function, from property (v) in Section 2.1.1, it follows that

$$\exp\left[\ln\left(w_1 x_1 + w_2 x_2 + \ldots + w_n x_n\right)\right] \geq \exp\left[\ln\left(x_1^{w_1} x_2^{w_2} \ldots x_n^{w_n}\right)\right]$$

or finally

$$w_1 x_1 + w_2 x_2 + \ldots + w_n x_n \geq x_1^{w_1} x_2^{w_2} \ldots x_n^{w_n} \tag{2.10}$$

Raising both sides of inequality (2.10) to the positive power of M, according to property (v) in Section 2.1.1, gives inequality (2.9):

$$\left(\frac{m_1 x_1 + m_2 x_2 + \ldots + m_n x_n}{M}\right)^M \geq \left(x_1^{m_1/M} x_2^{m_2/M} \ldots x_n^{m_n/M}\right)^M = x_1^{m_1} x_2^{m_2} \ldots x_n^{m_n}$$

2.1.4   ROOT-MEAN SQUARE – ARITHMETIC MEAN – GEOMETRIC MEAN – HARMONIC MEAN (RMS-AM-GM-HM) INEQUALITY

For a set of positive real numbers x_1, x_2, ..., x_n, the basic RMS-AM-GM-HM inequality states

$$\sqrt{\frac{x_1^2 + x_2^2 + \ldots + x_n^2}{n}} \geq \frac{x_1 + x_2 + \ldots + x_n}{n} \geq \sqrt[n]{x_1 x_2 \ldots x_n} \geq \frac{n}{1/x_1 + 1/x_2 + \ldots + 1/x_n} \tag{2.11}$$

with equality attained only if $x_1 = x_2 = \ldots = x_n$.

The number $\sqrt{\dfrac{x_1^2 + x_2^2 + \ldots + x_n^2}{n}}$ is known as *root-mean square*; $\dfrac{x_1 + x_2 + \ldots + x_n}{n}$ is the *arithmetic mean*; $\sqrt[n]{x_1 x_2 \ldots x_n}$ is the *geometric mean* and $\dfrac{n}{1/x_1 + 1/x_2 + \ldots + 1/x_n}$ is the *harmonic mean*.

For two non-negative numbers a, b, the inequality takes the form

$$\sqrt{\frac{a^2 + b^2}{2}} \geq \frac{a+b}{2} \geq \sqrt{ab} \geq \frac{2}{\dfrac{1}{a} + \dfrac{1}{b}} \tag{2.12}$$

Various techniques for proving the RMS-AM-GM-HM inequalities have been demonstrated in (Todinov, 2020a). The AM-GM inequality, for example, can be obtained as a special case of inequality (2.9) for $m_1 = m_2 = \ldots = m_n = 1$ $\left(M = \sum_{i=1}^{n} m_i = n \right)$. As a result,

$$\left(\frac{m_1 x_1 + m_2 x_2 + \ldots + m_n x_n}{M} \right)^M = \left(\frac{x_1 + x_2 + \ldots + x_n}{n} \right)^n \geq x_1^{m_1} x_2^{m_2} \ldots x_n^{m_n}$$

$$= x_1 x_2 \ldots x_n \qquad (2.13)$$

or finally

$$\frac{x_1 + x_2 + \ldots + x_n}{n} \geq \sqrt[n]{x_1 x_2 \ldots x_n} \qquad (2.14)$$

which completes the proof of the AM-GM inequality.

Segmentation through the AM-GM inequality (2.14) is a powerful technique for proving algebraic inequalities. The basic idea of this technique is to segment (split) the original inequality into simpler inequalities by using the AM-GM inequality and to sum the segmented inequalities in order to assemble the original inequality. Here is an example of this technique demonstrated on the algebraic inequality

$$x^2 + y^2 + z^2 \geq xy + yz + zx \qquad (2.15)$$

valid for any real x, y, z. The standard AM-GM inequality provides the following three inequalities:

$$\frac{y^2 + x^2}{2} \geq yx; \quad \frac{y^2 + z^2}{2} \geq yz; \quad \frac{z^2 + x^2}{2} \geq zx$$

which effectively segment the original inequality (2.15). Adding these three segments (inequalities) gives the original inequality (2.15) and completes the proof.

2.1.5   REARRANGEMENT INEQUALITY

The rearrangement inequality is a powerful yet underused basic algebraic inequality that can be applied for proving other inequalities.

Consider the two sequences a_1, a_2, $\ldots$, a_n and b_1, b_2, $\ldots$, b_n of real numbers. It can be shown that:

a.   The sum $S = a_1 b_1 + a_2 b_2 + \ldots + a_n b_n$ is maximal if the sequences are sorted in the same way: both monotonically decreasing: $a_1 \geq a_2 \geq, \ldots, \geq a_n$; $b_1 \geq b_2 \geq, \ldots, \geq b_n$ or both monotonically increasing: $a_1 \leq a_2 \leq, \ldots, \leq a_n$; $b_1 \leq b_2 \leq, \ldots, \leq b_n$.

b. The sum $S = a_1b_1 + a_2b_2 + \ldots + a_nb_n$ is minimal if the sequences are sorted in the opposite way: one monotonically increasing and the other monotonically decreasing. A proof of the rearrangement inequality can be found in (Todinov, 2020a).

The rearrangement inequality is a basis for generating new powerful inequalities that can be used to produce bounds for the uncertainty in reliability-critical parameters. For two sequences $a_1, a_2, \ldots, a_n$ and $b_1, b_2, \ldots, b_n$ of real numbers, the notation

$$\begin{bmatrix} a_1 & a_2 & \ldots & a_n \\ b_1 & b_2 & \ldots & b_n \end{bmatrix} = a_1b_1 + a_2b_2 + \ldots + a_nb_n$$

is introduced. This is similar to the definition of a dot product of two vectors with components specified by the two rows of the matrix. An important corollary of the rearrangement inequality is the following:

• Given a set of real numbers $a_1, a_2, \ldots, a_n$, for any permutation $a_{1s}, a_{2s}, \ldots, a_{ns}$, the following inequality holds:

$$a_1^2 + a_2^2 + \ldots + a_n^2 \geq a_1a_{1s} + a_2a_{2s} + \ldots + a_{ns} \qquad (2.16)$$

Indeed, without loss of generality, it can be assumed that $a_1 \leq a_2 \leq, \ldots, \leq a_n$. Applying the rearrangement inequality to the sequences $(a_1, a_2, \ldots, a_n), (a_1, a_2, \ldots, a_n)$ and to the sequences $(a_1, a_2, \ldots, a_n), (a_{1s}, a_{2s}, \ldots, a_{ns})$ then yields

$$\begin{bmatrix} a_1 & a_2 & \ldots & a_n \\ a_1 & a_2 & \ldots & a_n \end{bmatrix} = a_1^2 + a_2^2 + \ldots + a_n^2 \geq \begin{bmatrix} a_1 & a_2 & \ldots & a_n \\ a_{1s} & a_{2s} & \ldots & a_{ns} \end{bmatrix}$$

$$= a_1a_{1s} + a_2a_{2s} + \ldots + a_na_{ns}$$

because the first pair of sequences are similarly ordered and the second pair of sequences are not. This completes the proof of Inequality (2.16).

2.1.6 CHEBYSHEV'S SUM INEQUALITY

Another important basic algebraic inequality is the Chebyshev's sum inequality (Besenyei, 2018). It states that for the sequences of real numbers $a_1 \geq a_2 \geq, \ldots, \geq a_n$ and $b_1 \geq b_2 \geq, \ldots, \geq b_n$, the following inequality holds:

$$n\left(a_1b_1 + a_2b_2 + \ldots + a_nb_n\right) \geq \left(a_1 + a_2 + \ldots + a_n\right)\left(b_1 + b_2 + \ldots + b_n\right)$$

or

$$\frac{a_1b_1 + \ldots + a_nb_n}{n} \geq \frac{a_1 + \ldots + a_n}{n} \cdot \frac{b_1 + \ldots + b_n}{n} \tag{2.17}$$

If $a_1 \geq, \ldots, \geq a_n$ and $b_1 \leq, \ldots, \leq b_n$ hold, the inequality is reversed:

$$\frac{a_1b_1 + \ldots + a_nb_n}{n} \leq \frac{a_1 + \ldots + a_n}{n} \cdot \frac{b_1 + \ldots + b_n}{n} \tag{2.18}$$

Equality is attained if $a_1 = a_2 = \ldots = a_n$ or $b_1 = b_2 = \ldots = b_n$ holds.

For the sequences of real numbers $a_1 \geq a_2 \geq, \ldots, \geq a_n$ and $b_1 \geq b_2 \geq, \ldots, \geq b_n$, the Chebyshev's sum inequality (2.17) can be proved by using the rearrangement inequality discussed in the previous section. According to the rearrangement inequality, the following inequalities are true:

$$\sum_{i=1}^{n} a_ib_i = a_1b_1 + a_2b_2 + a_3b_3 + \ldots + a_nb_n$$

$$\sum_{i=1}^{n} a_ib_i \geq a_1b_2 + a_2b_3 + a_3b_4 + \ldots + a_nb_1$$

$$\sum_{i=1}^{n} a_ib_i \geq a_1b_3 + a_2b_4 + a_3b_5 + \ldots + a_nb_2$$

$$\sum_{i=1}^{n} a_ib_i \geq a_1b_n + a_2b_1 + a_3b_2 + \ldots + a_nb_{n-1}$$

By adding these inequalities, the inequality

$$n\sum_{i=1}^{n} a_ib_i \geq a_1\left(b_1 + b_2 + \ldots + b_n\right) + a_2\left(b_1 + b_2 + \ldots + b_n\right) + \ldots + a_n\left(b_1 + b_2 + \ldots + b_n\right)$$

is obtained, which, after factoring out the common term $(b_1 + b_2 + \ldots + b_n)$ leads to the Chebyshev's inequality (2.17). For sequences $a_1 \geq, \ldots, \geq a_n$ and $b_1 \leq, \ldots, \leq b_n$, the Chebyshev's sum inequality (2.18) can be proved in a similar fashion, by using the rearrangement inequality.

Chebyshev's sum inequality provides the unique opportunity to segment an initial complex problem into simple problems. The complex terms a_ib_i in inequalities (2.17) and (2.18) are segmented into simpler terms involving a_i and b_i.

The segmentation capability provided by the Chebyshev's sum inequality will be illustrated by evaluating the lower bound of $\sum_{i=1}^{n} x_i^2$ if $x_1 + x_2 + \ldots + x_n = 1$.

Without loss of generality, it can be assumed that $x_1 \geq x_2 \geq \ldots \geq x_n$. By setting $a_1 = x_1, a_2 = x_2, \ldots, a_n = x_n$ and $b_1 = x_1, b_2 = x_2, \ldots, b_n = x_n$, the conditions for applying the Chebyshev's inequality (2.17) are fulfilled and

$$\frac{x_1^2 + \ldots + x_n^2}{n} \geq \frac{x_1 + \ldots + x_n}{n} \cdot \frac{x_1 + \ldots + x_n}{n} \tag{2.19}$$

Substituting $x_1 + x_2 + \ldots + x_n = 1$ in (2.19) gives the lower bound of $\sum_{i=1}^{n} x_i^2$:

$$\sum_{i=1}^{n} x_i^2 \geq 1/n \tag{2.20}$$

2.1.7 MUIRHEAD'S INEQUALITY

Consider the two non-increasing sequences $a_1 \geq a_2 \geq, \ldots, \geq a_n$ and $b_1 \geq b_2 \geq, \ldots, \geq b_n$ of positive real numbers. The sequence $\{a\}$ is said to majorize the sequence $\{b\}$ if the following conditions are fulfilled:

$$a_1 \geq b_1; a_1 + a_2 \geq b_1 + b_2; \ldots; a_1 + a_2 + \ldots + a_{n-1} \geq b_1 + b_2 + \ldots + b_{n-1};$$

$$a_1 + a_2 + \ldots + a_{n-1} + a_n = b_1 + b_2 + \ldots + b_{n-1} + b_n \tag{2.21}$$

If the sequence $\{a\}$ majorizes the sequence $\{b\}$ and $x_1, x_2, \ldots, x_n$ are non-negative, the *Muirhead's inequality*

$$\sum_{sym} x_1^{a_1} x_2^{a_2} \ldots x_n^{a_n} \geq \sum_{sym} x_1^{b_1} x_2^{b_2} \ldots x_n^{b_n} \tag{2.22}$$

holds (Hardy et al., 1999).

For any set of non-negative numbers $x_1, x_2, \ldots, x_n$, the symmetric sum $\sum_{sym} x_1^{a_1} x_2^{a_2} \ldots x_n^{a_n}$, when expanded, includes $n!$ terms. Each term is formed by a permutation of the elements of the sequence $a_1, a_2, \ldots, a_n$. For example, if $\{a\} = [2, 1, 0]$ then

$$\sum_{sym} x_1^2 x_2^1 x_3^0 = x_1^2 x_2 + x_1^2 x_3 + x_2^2 x_1 + x_2^2 x_3 + x_3^2 x_1 + x_3^2 x_2$$

If $\{a\} = [2, 0, 0]$, then

$$\sum_{sym} x_1^2 x_2^0 x_3^0 = 2x_1^2 + 2x_2^2 + 2x_3^2$$

Here is an application example featuring an inequality that follows directly from the Muirhead's inequality (2.22). Consider a set of real, non-negative numbers x_1, x_2, x_3. It can be shown that the next inequality holds:

$$x_1^4 + x_2^4 + x_3^4 \geq x_1^2 x_2 x_3 + x_2^2 x_3 x_1 + x_3^2 x_1 x_2 \tag{2.23}$$

Consider the set of non-negative numbers x_1, x_2, x_3 and the sequences $\{a\} = [4, 0, 0]$ and $\{b\} = [2, 1, 1]$. Clearly, the sequence $\{a\} = [4, 0, 0]$ majorizes the sequence $\{b\} = [2, 1, 1]$ because the conditions (2.21) are fulfilled:

$$4 > 2; \; 4 + 0 > 2 + 1 \text{ and } 4 + 0 + 0 = 2 + 1 + 1.$$

According to the Muirhead's inequality (2.22):

$$2 \times \left(x_1^4 + x_2^4 + x_3^4 \right) \geq 2 \left(x_1^2 x_2 x_3 + x_2^2 x_3 x_1 + x_3^2 x_1 x_2 \right)$$

which implies inequality (2.23).

2.2   ALGEBRAIC INEQUALITIES THAT PERMIT NATURAL REVERSE ENGINEERING

2.2.1 SYMMETRIC ALGEBRAIC INEQUALITIES WHOSE TERMS CAN BE INTERPRETED AS PROBABILITIES

Some symmetric algebraic inequalities permit natural interpretation as probabilities if particular constraints are imposed on the variables of the inequalities. Such is for example the wide class of symmetric algebraic inequalities with imposed additional constraints. For example, the inequalities

$$x^2 + y^2 + z^2 \geq 1/3 \tag{2.24}$$

$$2xy + 2yz + 2zx \leq 2/3 \tag{2.25}$$

with imposed constraints $0 < x < 1$, $0 < y < 1$, $0 < z < 1$ and $x + y + z = 1$ admit natural interpretation of x, y and z as fractions of items of varieties X, Y and Z. Inequality (2.24) is a special case of inequality (2.20) for $n = 3$. Inequality (2.25) can be obtained from the identity $(x + y + z)^2 = x^2 + y^2 + z^2 + 2xy + 2yz + 2zx = 1$ and inequality (2.24). Indeed,

$$2xy + 2yz + 2zx = 1 - \left(x^2 + y^2 + z^2 \right) \leq 1 - 1/3 = 2/3$$

The left-hand side of inequality (2.24) is a sum of three terms which implies the application of *the total probability theorem* (DeGroot, 1989) in determining the probability of a compound event which includes three mutually exclusive events. Such a compound event for inequality (2.24) is, for example, selecting two items from the same variety from a large batch containing items of three varieties X, Y and Z. The corresponding fractions of items in the batch from the separate varieties are x, y and z ($0 < x < 1$, $0 < y < 1$, $0 < z < 1$ and $x + y + z = 1$). Selecting two items from variety X, variety Y or variety Z are mutually exclusive events, and according to the total probability theorem, their probabilities are added to obtain the probability of selecting two items of the same variety.

Similarly, the left side of inequality (2.25) can be interpreted as the probability of selecting two items from two different varieties. Selecting two items of varieties XY, YZ or ZX are mutually exclusive events and, according to the total probability theorem, their probabilities $2xy$, $2yz$ and $2zx$ are added.

Symmetric inequalities similar to

$$(1-a)(1-b)(1-c) \le 1 - abc \qquad (2.26)$$

where $0 \le a \le 1$, $0 \le b \le 1$, $0 \le c \le 1$, can be reverse engineered easily if a, b and c are chosen to denote the reliabilities of components A, B, and C, correspondingly. The reliabilities a, b and c are effectively the probabilities that the components will be in working condition. The probabilities that the components will be in a 'failed condition' are $1 - a$, $1 - b$ and $1 - c$, correspondingly.

The system in Figure 2.2a, where the components A, B and C have been logically arranged in parallel, is in a failed state only if all components are in a failed state, while the system in Figure 2.2b, where all components are logically arranged in series, is in a working state only if all components are working. Consequently, the left hand-side of inequality (2.26) can be interpreted as probability of failure of the system in Figure 2.2a and the right-hand side of the inequality can be interpreted as probability of failure of the system in Figure 2.2b.

The reverse engineering of inequality (2.26) yields that the probability of failure of the system in Figure 2.2a is smaller than the probability of failure of the system in Figure 2.2b.

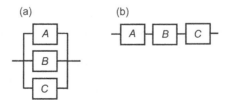

FIGURE 2.2 Three components A, B and C logically arranged (a) in parallel and (b) in series.

2.2.2 TRANSFORMING ALGEBRAIC INEQUALITIES TO MAKE THEM SUITABLE FOR REVERSE ENGINEERING

By appropriate transformations, some symmetric inequalities can be made suitable for reverse engineering. A common transformation technique is multiplying both sides of the inequality by a constant. Thus, multiplying both sides by the constant (1/2), for example, creates the possibility for interpretation terms of type a^2/b as elastic strain energy, electric power or stored electric energy (see Chapter 3 for details). Multiplying both sides of an inequality by $1/n$ where n is the number of terms in the left-hand side of the inequality, creates the possibility to apply the total probability theorem and interpret the inequality as a probability of a compound event composed of mutually exclusive events.

Indeed, if event B can occur with any of the n mutually exclusive and exhaustive events A_i $\left(P\left(A_i \cap A_j \right) = 0 \text{ and } \sum_{i=1}^{n} P\left(A_i \right) = 1 \right)$, according to the total probability theorem, the probability $P(B)$ of event B is given by:

$$P\left(B \right) = P(B \mid A_1)P\left(A_1 \right) + P(B \mid A_2)P\left(A_2 \right) + ... + P(B \mid A_n)P\left(A_n \right) \quad (2.27)$$

Now if events A_i have the same chance of occurring, the probability $P(A_i)$ of their occurrence is equal to $1/n$. The probability $P(B)$ of event B then becomes

$$P\left(B \right) = \left(1/n \right)P(B \mid A_1) + \left(1/n \right)P(B \mid A_2) + ... + \left(1/n \right)P(B \mid A_n) \quad (2.28)$$

Inequality (2.15) that has been proved in Section 2.1.4 is one such example. Multiplying both sides of inequality (2.15) by 1/3 gives the inequality

$$\left(1/3 \right)x^2 + \left(1/3 \right)y^2 + \left(1/3 \right)z^2 \geq \left(1/3 \right)xy + \left(1/3 \right)yz + \left(1/3 \right)zx \quad (2.29)$$

If x, y and z stand for the fractions of reliable components characterising three suppliers X, Y and Z, the left side of inequality (2.29) can be interpreted as probability of purchasing two reliable components from a randomly selected supplier. Indeed, the term, $(1/3)x^2$ in inequality (2.29) is the probability that the first supplier will be randomly selected and both components purchased from this supplier will be reliable. The term, $(1/3)y^2$ is the probability that the second supplier will be randomly selected and both components purchased from this supplier will be reliable. Finally, the term, $(1/3)z^2$ is the probability that the third supplier will be randomly selected and both components purchased from this supplier will be reliable. These events are mutually exclusive and exhaustive and, according to the total probability theorem, $p_1 = (1/3)x^2 + (1/3)y^2 + (1/3)z^2$ is the probability of purchasing two reliable components from a randomly selected supplier. With a similar reasoning, it can be shown that the right-hand side of inequality (2.29)

$p_2 = (1/3)xy + (1/3)yz + (1/3)zx$ can be interpreted as probability of purchasing two reliable components from two randomly selected suppliers.

The next example of transformations applied to symmetric inequalities is related to the inequality

$$a^2 + b^2 \geq 2ab \tag{2.30}$$

where $a, b > 0$. The transformation consists of dividing both sides of the inequality by the positive quantity $a + b$. This operation does not alter the direction of the inequality but the terms $x = a/(a + b)$ and $y = b/(a + b)$ can be physically interpreted as fractions of items from two varieties X and Y, correspondingly, in a large batch of items. The original inequality is reduced to

$$x^2 + y^2 \geq 2xy \tag{2.31}$$

where $x, y > 0$ and $x + y = 1$. The left-hand side of inequality (2.31) can now be interpreted as the probability of selecting two items of the same variety from a large batch and the right-hand side can be interpreted as the probability of selecting two items of different variety from the same batch.

2.2.3 INEQUALITIES BASED ON SUB-ADDITIVE AND SUPER-ADDITIVE FUNCTIONS

This is a very important class of algebraic inequalities that provides the possibility for segmentation of a controlling factor. Inequalities based on sub-additive and super-additive functions have a number of key potential applications in enhancing the effect of additive quantities (factors).

Extensive quantities are typical additive quantities. They change with changing the size of their supporting objects/systems (DeVoe, 2012; Mannaerts, 2014). DeVoe (2012), for example, introduced 'extensivity' test which consists of dividing the system by an imaginary surface into two parts. Any quantity characterising the system that is the sum of the same quantity characterising the two parts is an *extensive* quantity and any quantity that has the same value in each part of the system is an *intensive* quantity.

Examples of extensive quantities are mass, weight, amount of substance, number of particles, volume, distance, energy (kinetic energy, gravitational energy, electric energy, elastic energy, surface energy, internal energy), work, power, heat, force, momentum, electric charge, electric current, heat capacity, electric capacity, resistance (when the elements are in series), enthalpy and fluid flow.

Intensive quantities characterise the object/system locally and do not change with changing the size of the supporting system. Examples of intensive quantities are: temperature, pressure, density, concentration, hardness, velocity and surface tension.

In the Euclidean space of one dimension, the sub-additive function $f(x)$ satisfies the inequality (Alsina and Nelsen, 2010):

$$f\left(x_1 + x_2\right) \le f\left(x_1\right) + f\left(x_2\right) \qquad (2.32)$$

for any pair of points x_1 and x_2 in the domain of definition. If the direction of the inequality is reversed, the function $f(x)$ is super-additive:

$$f\left(x_1 + x_2\right) \ge f\left(x_1\right) + f\left(x_2\right) \qquad (2.33)$$

In the Euclidean space of two dimensions, the multivariable sub-additive function $f(x, y)$ satisfies the inequality

$$f\left(x_1 + x_2, y_1 + y_2\right) \le f\left(x_1, y_1\right) + f\left(x_2, y_2\right) \qquad (2.34)$$

for any pair of points (x_1, y_1) and (x_2, y_2) in the domain of definition. If the direction of the inequality is reversed, the function $f(x, y)$ is super-additive:

$$f\left(x_1 + x_2, y_1 + y_2\right) \ge f\left(x_1, y_1\right) + f\left(x_2, y_2\right) \qquad (2.35)$$

Consider n points with positive coordinates x_i in the definition domain of the function $f()$. By induction, from definition (2.32), it follows that

$$f\left(x_1 + x_2 + \ldots + x_n\right) \le f\left(x_1\right) + f\left(x_2\right) + \ldots + f\left(x_n\right) \qquad (2.36)$$

while from definition (2.33), it follows that

$$f\left(x_1 + x_2 + \ldots + x_n\right) \ge f\left(x_1\right) + f\left(x_2\right) + \ldots + f\left(x_n\right) \qquad (2.37)$$

Key results related to sub-additive and super-additive functions of a single variable have been stated in (Alsina and Nelsen, 2010). Thus, if a function $f(x)$, with a domain $[0, \infty)$ and range $[0, \infty)$ is concave, and $f(0) \ge 0$ then the function is sub-additive: $f(x_1 + x_2 + \ldots + x_n) \le f(x_1) + f(x_2) + \ldots + f(x_n)$. If the function $f(x)$ is convex and $f(0) \le 0$, then it is super-additive: $f(x_1 + x_2 + \ldots + x_n) \ge f(x_1) + f(x_2) + \ldots + f(x_n)$.

Consider now n pairs of points (a_i, b_i) with positive coordinates a_i, b_i from the definition domain of the function $f(x, y)$. From the definition (2.34), it follows that

$$f\left(a_1 + a_2 + \ldots + a_n, b_1 + b_2 + \ldots + b_n\right) \le f\left(a_1, b_1\right) + f\left(a_2, b_2\right) + \ldots + f\left(a_n, b_n\right) \quad (2.38)$$

while from the definition (2.35), it follows that

$$f\left(a_1 + a_2 + \ldots + a_n, b_1 + b_2 + \ldots + b_n\right) \ge f\left(a_1, b_1\right) + f\left(a_2, b_2\right) + \ldots + f\left(a_n, b_n\right) \quad (2.39)$$

Relationships (2.36)–(2.39) can be obtained easily by mathematical induction and derivation details have been omitted.

Inequalities (2.38) and (2.39) do not change their direction upon any permutation of a_i and b_i. They have a number of powerful potential applications in increasing/decreasing the effect of additive quantities (factors).

Inequalities (2.36) and (2.37) can, for example, be applied for generating new knowledge that can be used for optimising processes. If the function $f(x)$ describes the effect/output of an additive controlling factor x and x_i $(i = 1, ..., n)$ denotes a smaller part of the factor, inequalities (2.36) and (2.37) provide the unique opportunity to increase the effect of the factor, by segmenting it or aggregating it, depending on whether the function $f(x)$ is concave or convex. If the function is concave, with a domain $[0, \infty)$ and range $[0, \infty)$, segmenting the factor results in a larger output. Conversely, if the function is convex in this domain and if the function is zero when the factor is zero, aggregating the factor results in a larger output.

An important condition for using inequalities (2.36) and (2.37) is the outputs $f(x_1), f(x_2), ..., f(x_n)$ to be also additive, after segmenting the additive controlling factor x. This is fulfilled if the outputs $f(x_1), f(x_2), ..., f(x_n)$ are additive quantities such as volume, surface area, energy, power, force, mass, damage, number and amount of profit.

It is important to note that not all abstract algebraic inequalities have a direct physical interpretation. Consider the general inequality

$$\varphi_1\left(x, y, z\right) + \varphi_2\left(x, y, z\right) + \varphi_3\left(x, y, z\right) > \psi\left(x, y, z\right)$$

correct for any positive x, y and z. In this inequality, $\varphi_i()$ and $\psi()$ are particular functions of the variables x, y and z.

Suppose that the inequality is interpreted as a physical system or process and the variables x, y, and z have a particular physical interpretation.

To make sense for a physical system or process all terms $\varphi_i(x, y, z)$ in the left-hand side must have the same physical meaning and the same units of measurement, otherwise the terms $\varphi_i(x, y, z)$ in the left-hand side simply cannot be added together. In addition, the quantity $\psi(x, y, z)$ in the right-hand side, must also have the same physical meaning and the same units of measurement as the left-hand side. Alternatively, all additive terms of the inequality must be dimensionless and have the same physical meaning.

For a number of correct abstract algebraic inequalities, these requirements are too restrictive and for this reason, no physical meaning can be attached to such inequalities.

In order to apply inequalities (2.38) or (2.39), a_i, b_i and the output quantities $f(a_i, b_i)$ must all represent additive quantities. Inequality (2.38) effectively states that the effect of the additive quantities $a = \sum_{i=1}^{n} a_i$ and $b = \sum_{i=1}^{n} b_i$ can be increased by segmenting them into smaller parts a_i and b_i, $i = 1, ..., n$ and accumulating their effects $f(a_i, b_i)$, represented by the sum of the terms on the right-hand side of inequality (2.38). Similarly, inequality (2.39) effectively states that the effect of the additive quantities $a = \sum_{i=1}^{n} a_i$ and $b = \sum_{i=1}^{n} b_i$ can be decreased by segmenting them into smaller parts a_i and b_i, $i = 1, \dots, n$ and accumulating their effects $f(a_i, b_i)$, represented by the sum of the terms on the right-hand side of inequality (2.39).

Inequalities (2.38) and (2.39) have a universal application in science and technology as long as a_i, b_i and the terms $f(a_i, b_i)$ are additive quantities and have the same physical interpretation.

Inequalities similar to (2.38) and (2.39) can be derived for any number of factors by using the definition of sub-/super-additive functions. For example, for three factors a, b and c, the inequality

$$f\left(a_1 + \dots + a_n, b_1 + \dots + b_n, c_1 + \dots + c_n\right) \le f\left(a_1, b_1, c_1\right) + f\left(a_2, b_2, c_2\right) + \dots$$
$$+ f\left(a_n, b_n, c_n\right) \tag{2.40}$$

is applicable for a sub-additive function $f(a, b, c)$ and the inequality

$$f\left(a_1 + \dots + a_n, b_1 + \dots + b_n, c_1 + \dots + c_n\right) \ge f\left(a_1, b_1, c_1\right) + f\left(a_2, b_2, c_2\right) + \dots$$
$$+ f\left(a_n, b_n, c_n\right) \tag{2.41}$$

is applicable for a super-additive function $f(a, b, c)$.

Inequalities based on sub-additive and super-additive functions are important classes of algebraic inequalities that permit easy reverse engineering. The knowledge derived from the reverse engineering of these inequalities can be used to achieve superior performance from a system or process.

Sufficient conditions for sub-additivity and super-additivity of a single-variable function can also be stated. If a function $f(x)$, with a domain $[0, \infty)$ and range $[0, \infty)$ is a concave single-variable function (see Section 2.13 for a definition), then the function is sub-additive $(f(x_1 + x_2 + \dots + x_n) \le f(x_1) + f(x_2) + \dots + f(x_n))$. If the function $f(x)$ is convex, and if $f(0) \le 0$, then the function is super-additive $(f(x_1 + x_2 + \dots + x_n) \ge f(x_1) + f(x_2) + \dots + f(x_n))$.

Proof: The correctness of the first statement can be proved by an argument based on the fact that $f(x)$ is a concave function. Note that $w_{1,k} = x_k / \sum_{i=1}^{n} x_i$ and

$$w_{2,k} = \left(\sum_{i=1}^{n} x_i - x_k\right) / \sum_{i=1}^{n} x_i \text{ can be treated as weights because } 0 \le w_{1,k} \le 1, 0$$

$\le w_{2,k} \le 1$ and $w_{1,k} + w_{2,k} = 1$. Because $f(x)$ is a concave function, for the values $x = x_1 + x_2 + \ldots + x_n$ and $x = 0$, the Jensen's inequality for concave functions gives:

$$f(x_1) = f\left(\frac{x_1}{\sum\limits_{i=1}^{n} x_i} \times \sum\limits_{i=1}^{n} x_i + \frac{\sum\limits_{i=1}^{n} x_i - x_1}{\sum\limits_{i=1}^{n} x_i} \times 0\right) \geq \frac{x_1}{\sum\limits_{i=1}^{n} x_i} f(x_1 + \ldots + x_n)$$

$$+ \frac{\sum\limits_{i=1}^{n} x_i - x_1}{\sum\limits_{i=1}^{n} x_i} f(0) \qquad (2.42)$$

$$f(x_2) = f\left(\frac{x_2}{\sum\limits_{i=1}^{n} x_i} \times \sum\limits_{i=1}^{n} x_i + \frac{\sum\limits_{i=1}^{n} x_i - x_2}{\sum\limits_{i=1}^{n} x_i} \times 0\right) \geq \frac{x_2}{\sum\limits_{i=1}^{n} x_i} f(x_1 + \ldots + x_n)$$

$$+ \frac{\sum\limits_{i=1}^{n} x_i - x_2}{\sum\limits_{i=1}^{n} x_i} f(0) \qquad (2.43)$$

. . . .

$$f(x_n) = f\left(\frac{x_n}{\sum\limits_{i=1}^{n} x_i} \times \sum\limits_{i=1}^{n} x_i + \frac{\sum\limits_{i=1}^{n} x_i - x_n}{\sum\limits_{i=1}^{n} x_i} \times 0\right) \geq \frac{x_n}{\sum\limits_{i=1}^{n} x_i} f(x_1 + \ldots + x_n)$$

$$+ \frac{\sum\limits_{i=1}^{n} x_i - x_n}{\sum\limits_{i=1}^{n} x_i} f(0) \qquad (2.44)$$

Adding all n inequalities (2.42)–(2.44) yields

$$f(x_1) + f(x_2) + \ldots + f(x_n) \geq f(x_1 + x_2 + \ldots + x_n) + (n-1)f(0) \qquad (2.45)$$

Since $f(0) \geq 0$ (the range of the function $f(x)$ is $[0, \infty)$), we have

$$f(x_1) + f(x_2) + \ldots + f(x_n) \geq f(x_1 + x_2 + \ldots + x_n)$$
$$+ (n-1)f(0) \geq f(x_1 + x_2 + \ldots + x_n)$$

which completes the proof of inequality (2.36).

Note that if $f(0) > 0$ the following strict inequality can be obtained from inequality (2.45):

$$f(x_1) + f(x_2) + \ldots + f(x_n) \geq f(x_1 + x_2 + \ldots + x_n) + (n-1)f(0)$$
$$> f(x_1 + x_2 + \ldots + x_n)$$

In the same way, inequality (2.37) can also be proved if the function $f(x)$ is convex and $f(x) \leq 0$. These are sufficient conditions for super-additivity of a single-variable function.

Inequalities can also be based on multivariable sub-additive and super-additive functions.

Multivariate sub-additive functions have been discussed in Rosenbaum (1950). A special case of the general sub-additive function (2.39) is the algebraic inequality

$$\sqrt{ab} \geq \sqrt{a_1 b_1} + \sqrt{a_2 b_2} + \ldots + \sqrt{a_n b_n} \tag{2.46}$$

where both controlling factors $a = a_1 + a_2 + \ldots + a_n$ and $b = b_1 + b_2 + \ldots + b_n$ are additive quantities. The role of the function $f(a, b)$ in inequality (2.39) is played by the function $f(a,b) \equiv \sqrt{ab}$ in inequality (2.46).

Inequality (2.46) can be derived from the Cauchy-Schwarz inequality by making the substitutions: $x_1 = \sqrt{a_1}$, $x_2 = \sqrt{a_2}, \ldots$, $x_n = \sqrt{a_n}$; $y_1 = \sqrt{b_1}$, $y_2 = \sqrt{b_2}$ $, \ldots, y_n = \sqrt{b_n}$. Applying the Cauchy-Schwarz inequality to the sequences $x_1, x_2, \ldots, x_n$ and $y_1, y_2, \ldots, y_n$, yields inequality (2.46):

$$\sqrt{a_1 b_1} + \sqrt{a_2 b_2} + \ldots + \sqrt{a_n b_n} \leq \sqrt{\left[\left(\sqrt{a_1}\right)^2 + \ldots + \left(\sqrt{a_n}\right)^2\right] \times \left[\left(\sqrt{b_1}\right)^2 + \ldots + \left(\sqrt{b_n}\right)^2\right]}$$

Physical meaning can be created for the quantities a_i, b_i entering inequality (2.46). The condition for reverse engineering of inequality (2.46) is the quantity represented by $\sqrt{a_i b_i}$ to be an additive quantity.

2.2.4 BERGSTRÖM INEQUALITY AND ITS NATURAL PHYSICAL INTERPRETATION

Often transformations of basic algebraic inequalities are a necessary prerequisite for their reverse engineering. Such a transformation will be demonstrated with the standard Cauchy-Schwarz inequality (2.1). If the substitutions $a_i = \dfrac{x_i}{\sqrt{y_i}}$ and $b_i = \sqrt{y_i}$ ($i = 1, \ldots, n$) are made in the Cauchy-Schwarz inequality (2.1):

$$\left(\frac{x_1}{\sqrt{y_1}}\sqrt{y_1} + \frac{x_2}{\sqrt{y_2}}\sqrt{y_2} + \ldots + \frac{x_n}{\sqrt{y_n}}\sqrt{y_n}\right)^2$$
$$\leq \left(\left(x_1/\sqrt{y_1}\right)^2 + \ldots + \left(x_n/\sqrt{y_n}\right)^2\right)\left(\left(\sqrt{y_1}\right)^2 + \ldots + \left(\sqrt{y_n}\right)^2\right)$$

the result is the Bergström inequality (Pop, 2009; Sedrakyan and Sedrakyan, 2010):

$$\frac{x_1^2}{y_1} + \frac{x_2^2}{y_2} + \ldots + \frac{x_n^2}{y_n} \geq \frac{\left(x_1 + x_2 + \ldots + x_n\right)^2}{y_1 + y_2 + \ldots + y_n} \tag{2.47}$$

which is effectively the modified Cauchy-Schwarz inequality. Inequality (2.47) is valid for any sequence $x_1, x_2, \ldots, x_n$ of real numbers and any sequence $y_1, y_2, \ldots, y_n$ of positive real numbers. Equality in (2.47) is attained only if $x_1/y_1 = x_2/y_2 = \ldots = x_n/y_n$.

In order to apply inequality (2.47) to additive quantities, the quantities x_i, y_i and the ratios x_i^2 / y_i must all be additive quantities. In this case, the inequality provides mechanism for increasing the effect of the quantities $x = \left(\sum_{i=1}^{n} x_i\right)^2$ and $y = \sum_{i=1}^{n} y_i$, by segmenting them into smaller quantities x_i and y_i, $i = 1, \ldots, n$ and accumulating the individual effects x_i^2 / y_i.

Various physical meaning can be created for the quantities x^2 and y entering inequality (2.47). For example, the quantity x^2 could be the square of force or voltage and the quantity b could be area, stiffness or electrical resistance. A necessary condition for physically interpreting inequality (2.47) is the quantity represented by x^2/y to be an additive quantity. Thus, the strain energy of a bar subjected to tension is proportional to Force^2/Area and the strain energy is an additive quantity. Similarly, Voltage^2/Resistance stands for dissipated power, which is also an additive quantity.

2.2.5 SUB-ADDITIVE AND SUPER-ADDITIVE INEQUALITIES BASED ON A CONCAVE AND CONVEX POWER LAW DEPENDENCE

Consider the sub-additive inequality

$$ax_1^p + ax_2^p + \ldots + ax_m^p < ay_1^p + ay_2^p + \ldots + ay_n^p \tag{2.48}$$

where $m < n$, $p < 1$, $x_i > 0$, $y_i > 0$, $a > 0$ and $\sum_{i=1}^{m} x_i = \sum_{i=1}^{n} y_i$. In general, inequality (2.48) holds under the following sufficient majorising conditions:

$$\begin{aligned}
x_1 &\geq y_1; x_1 \geq y_2; \ldots; x_1 \geq y_n \\
x_2 &\geq y_1; x_2 \geq y_2; \ldots; x_2 \geq y_n \\
x_m &\geq y_1; x_m \geq y_2; \ldots; x_m \geq y_n
\end{aligned} \tag{2.49}$$

The majorising conditions (2.49) effectively state that any x_i in the left-hand side of inequality (2.48) majorises each y_i in the right-hand side of the inequality. (In general, x_i and y_i are not necessarily equal). The proof of inequality (2.48) under these sufficient conditions has been given in Section 2.2.5.1.

For equal x_i and y_i ($x_1 = x_2 = \ldots = x_m = z/m$ and $y_1 = y_2 = \ldots = y_n = z/n$), the sufficient majorising conditions (2.49) are automatically satisfied and inequality (2.48) holds true. For this important special case however, a simplified proof of the inequality has been presented in Section 2.2.5.2.

For $p > 1, m < n, x_i > 0, y_i > 0, a > 0, \sum_{i=1}^{m} x_i = \sum_{i=1}^{n} y_i$ and under the same sufficient conditions (2.49) specified for inequality (2.48), it can be proved that the super-additive inequality

$$ax_1^p + ax_2^p + \ldots + ax_m^p > ay_1^p + ay_2^p + \ldots + ay_n^p \qquad (2.50)$$

holds true. (The details of the proof are similar to those related to inequality (2.48) and will be omitted here).

The primary advantage of inequalities (2.48) and (2.50) lies in their simplicity and ease of physical interpretation which renders them particularly suitable for reverse engineering. In inequalities (2.48) and (2.50), x and y can be interpreted as additive quantities while 'a' and 'p' are constants. For a meaningful physical interpretation, each individual term ax_i^p and ay_i^p within the inequalities must also be an additive quantity.

2.2.5.1 Proof of Inequality (2.48) in the General Case

The proof of inequality (2.48) will be made for unequal x_i and y_i.

Let the conditions (2.49) and the condition:

$$\sum_{i=1}^{m} x_i = \sum_{i=1}^{n} y_i \qquad (2.51)$$

be fulfilled.

Because $m < n$, it can be assumed that the left and right-hand side of (2.48) have the same number of n terms, where $x_{m+1} = x_{m+2} = \ldots = x_n = 0$. Without loss of generality it can be assumed that

$$x_1 \geq x_2 \geq \ldots \geq x_n \qquad (2.52)$$

$$y_1 \geq y_2 \geq \ldots \geq y_n \qquad (2.53)$$

are fulfilled. Considering 2.49, 2.52 and 2.53, the following majorisation property can be established:

$$\begin{aligned}
&x_1 \geq y_1;\ x_1 + x_2 \geq y_1 + y_2;\ldots; \\
&x_1 + x_2 + \ldots + x_{n-1} \geq y_1 + y_2 + \ldots + y_{n-1}; \\
&x_1 + x_2 + \ldots + x_n = y_1 + y_2 + \ldots + y_n;
\end{aligned} \qquad (2.54)$$

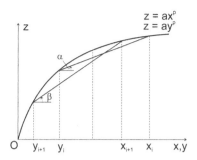

FIGURE 2.3 The key property (2.55) for the concave power law function $z = ax^p$ ($z = ay^p$).

Next, if for any i, $x_i = y_i$, the inequality (2.48) will not be affected if x_i and y_i are removed. As a result, without loss of generality, it can be assumed that $x_i \neq y_i$, for all i.

Consider the slope of the secant $\dfrac{ax^p - ay^p}{x - y}$ through the points (x, ax^p) and (y, ay^p). The power law function $z = ax^p$ ($z = ay^p$) is concave and strictly increasing function in x and considering also conditions (2.52) and (2.53), implies that the following property holds (see Figure 2.3):

$$k_i = \tan\left(\alpha\right) = \frac{ax_i^p - ay_i^p}{x_i - y_i} < \frac{ax_{i+1}^p - ay_{i+1}^p}{x_{i+1} - y_{i+1}} = k_{i+1} = \tan\left(\beta\right) \qquad (2.55)$$

Let $X_0 = Y_0 = 0$ and $X_i = x_1 + x_2 + \ldots + x_i$, $Y_i = y_1 + y_2 + \ldots + y_i$, $i = 1, \ldots, n$

From the majorisation property (2.54), it follows that $X_i \geq Y_i$ for $i = 1, \ldots, n - 1$ and $X_n = Y_n$. Proving inequality (2.48) is equivalent to proving the inequality $\displaystyle\sum_{i=1}^{n}\left[ax_i^p - ay_i^p\right] < 0$.

From (2.55), it follows that

$$\sum_{i=1}^{n}\left[ax_i^p - ay_i^p\right] = \sum_{i=1}^{n}k_i\left(x_i - y_i\right) \qquad (2.56)$$

Since $x_i = X_i - X_{i-1}$ and $y_i = Y_i - Y_{i-1}$, the sum in the right-hand side of (2.56) can be presented as:

$$\sum_{i=1}^{n}k_i\left(x_i - y_i\right) = \sum_{i=1}^{n}k_i\left[X_i - X_{i-1} - \left(Y_i - Y_{i-1}\right)\right] = \sum_{i=1}^{n}k_i\left(X_i - X_{i-1}\right)$$

$$- \sum_{i=1}^{n}k_i\left(Y_i - Y_{i-1}\right) \qquad (2.57)$$

In turn, the sum in the right-hand side of Equation (2.57) can be presented as:

$$\sum_{i=1}^{n} k_i\left(X_i - X_{i-1}\right) - \sum_{i=1}^{n} k_i\left(Y_i - Y_{i-1}\right) = k_1\left(X_1 - Y_1\right) + k_2\left(X_2 - Y_2\right) + k_3\left(X_3 - Y_3\right) + \ldots$$
$$+ k_n\left(X_n - Y_n\right) - k_1\left(X_0 - Y_0\right) - k_2\left(X_1 - Y_1\right) - k_3\left(X_2 - Y_2\right) - \ldots - k_n\left(X_{n-1} - Y_{n-1}\right)$$

As a result, the right-hand side of Equation (2.57) becomes:

$$\sum_{i=1}^{n} k_i\left(X_i - X_{i-1}\right) - \sum_{i=1}^{n} k_i\left(Y_i - Y_{i-1}\right) = k_n\left(X_n - Y_n\right) - k_1\left(X_0 - Y_0\right)$$
$$+ \sum_{i=1}^{n-1}\left(k_i - k_{i+1}\right)\left(X_i - Y_i\right)$$

Considering that $k_n(X_n - Y_n) = 0$, $k_1(X_0 - Y_0) = 0$, the right-hand side of Equation (2.57) becomes:

$$\sum_{i=1}^{n} k_i\left(X_i - X_{i-1}\right) - \sum_{i=1}^{n} k_i\left(Y_i - Y_{i-1}\right) = 0 - 0 + \sum_{i=1}^{n-1}\left(k_i - k_{i+1}\right)\left(X_i - Y_i\right)$$

As a result, the relationship $\sum_{i=1}^{n} k_i\left(x_i - y_i\right) = \sum_{i=1}^{n-1}\left(k_i - k_{i+1}\right)\left(X_i - Y_i\right)$ has been estab-

lished. Since $X_i - Y_i \geq 0$ and $k_i - k_{i+1} < 0$, it follows that $\sum_{i=1}^{n} k_i\left(x_i - y_i\right) < 0$, which

proves inequality (2.48).

The truth of inequality (2.50) can be established by a very similar reasoning.

2.2.5.2   Proof of Inequality (2.48) in the Special Case

Consider now the special case of equal x_i and y_i:

$$x_1 = x_2 = \ldots = x_m = z / m \text{ and } y_1 = y_2 = \ldots = y_n = z / n$$

$$\sum_{i=1}^{m} x_i = z, \ \sum_{i=1}^{m} y_i = z, \ 0 < p < 1$$

In this case, inequality (2.48) can be proved by substituting $x_i = z/m$ and $y_i = z/n$ which results in the equivalent inequality

$$m \times a\left(z / m\right)^p < n \times a\left(z / n\right)^p \qquad (2.58)$$

Proving the last inequality is equivalent to proving the inequality

$$az^p \left(n^{1-p} - m^{1-p} \right) > 0 \tag{2.59}$$

The left-hand side of inequality (2.59) is always positive because $n^{1-p}/m^{1-p} = (n/m)^{1-p}$ > 1 if $n > m$ and $0 < p < 1$. From this, it follows that if $m < n$ and $0 < p < 1$ then $n^{1-p} - m^{1-p} > 0$ which completes the proof of inequality (2.59) and the equivalent inequalities (2.58) and (2.48).

The truth of inequality (2.50) for equal x_i and y_i can be established by a very similar reasoning and details will be omitted.

2.3 TESTING ALGEBRAIC INEQUALITIES BY MONTE CARLO SIMULATION

Before attempting to prove a derived or conjectured inequality rigorously, it is important to confirm it first by testing. An attempt is made to prove the inequality rigorously only if the testing provides support for the conjectured inequality. No attempt is made to prove the inequality rigorously if, during the testing, the inequality has been falsified by a counterexample.

Suppose that n distinct components with unknown reliabilities $a_1, a_2, ..., a_n$ are used for building two system configurations, with reliabilities given by the functions $f(a_1, ..., a_n)$ and $g(a_1, ..., a_n)$. Consider a case where it is necessary to test the conjecture that the system configuration with reliability $f(a_1, ..., a_n)$ is intrinsically superior to the system configuration with reliability $g(a_1, ..., a_n)$, irrespective of the reliabilities $a_1, a_2, ..., a_n$ of the components building the systems. To perform this test, it suffices to test that the conjectured inequality $f(a_1, ..., a_n) > g(a_1, ..., a_n)$ is not contradicted during multiple simulation trials involving random combinations of values for the unknown reliabilities $a_1, a_2, ..., a_n$. This can be done by running the Monte Carlo simulation Algorithm 2.1, given in pseudo-code.

In general, the variables a_i entering the conjectured inequality $f(a_1, ..., a_n) > g(a_1, ..., a_n)$, vary within specified intervals $L_i \le a_i \le U_i$, $i = 1, ..., n$. If a_i are reliabilities of components then $L_i = 0$, $U_i = 1$.

The essence of Algorithm 2.1 for testing a conjectured inequality, is repeated sampling from the intervals of variation of each variable a_i entering the inequality, substituting the sampled values in the inequality and checking whether the inequality $f(a_1, ..., a_n) > g(a_1, ..., a_n)$ holds. Even a single combination of values for the variables a_i $(i = 1, ..., n)$, for which the inequality does not hold, disproves the inequality and shows that there is no point in attempting rigorous proof because a counterexample had been found which falsified the conjectured inequality.

If the inequality holds for millions of generated random values for the sampled variables, a strong support is obtained for the conjecture that the tested inequality is true. Such support however, cannot replace a rigorous proof. The inequality must still be proved rigorously by using some of the techniques for proving algebraic inequalities.

The algorithm in pseudo-code, for confirming or disproving an inequality by a Monte simulation is given next.

Algorithm 2.1

```
a=[];
L=[L1,L2,…,Ln]; U=[U1,U2,…,Un];

flag=0;
for k=1 to num_trials do
{
for k=1 to n do a[k]=L[k]+(U[k] - L[k]) x rand();

y1=f(a[1],a[2],…,a[n]);
y2=g(a[1],a[2],…,a[n]);
y=y1-y2;

if(y<0) then { flag=1; break;}

}

if (flag==0) then print('The tested inequality has
never been contradicted');
else print('The tested inequality has been disproved');
```

Initially, in the loop

```
for k=1 to n do a[k]=L[k]+(U[k]-L[k]) x rand(),
```

random values are assigned to the variables $a[k]$ $(k=1,…,n)$ entering the inequality. This is done by calling the function **rand**(), which returns a random number uniformly distributed in the interval (0,1). The value returned by **rand**() is transformed linearly by the statement $a[k]=L[k]+(U[k]-L[k])$ x **rand**() into a random value $a[k]$ uniformly distributed in the interval $(L[k],U[k])$. $L[k]$ and $U[k]$ are the lower and upper limit of the range for variable $a[k]$.

Next, the left- and right-hand sides of the tested inequality are evaluated by calling the functions $f()$ and $g()$ with the statements

```
y1=f(a[1],a[2],…,a[n]); y2=g(a[1],a[2],…,a[n]);
```

The values returned by the functions are stored in the variables $y1$ and $y2$.

The difference $y=y1-y2$ is then checked for being smaller than zero. A difference smaller than zero corresponds to a case where the tested inequality has been contradicted. In the statement

```
if(y<0) then { flag=1; break;}
```

If the tested inequality has been contradicted, a variable serving as a flag is assigned a value of one (flag=1), and the simulation loop is exited immediately with the statement `break` because a counterexample has been found. At the end of the algorithm, the content of the variable `flag` is checked. If the flag remained zero during the simulations, this is an indication that the tested inequality has never been contradicted during the simulation trials. This indicates that the tested inequality is probably true and a rigorous proof can be attempted. If, during the simulations, the variable `flag` changed its value to one, this is an indication that there had been a combination of values for the variables entering the tested inequality for which the inequality does not hold. This means that there is no point in searching for a rigorous proof because a counterexample has been found, which disproves the conjectured inequality.

Consider the non-trivial algebraic inequality:

$$\left(1-a^2b\right)\left(1-b^2c\right)\left(1-c^2a\right) \geq \left(1-a^3\right)\left(1-b^3\right)\left(1-c^3\right)$$

where a, b, c are positive real numbers smaller than 1. Before an attempt can be made to prove this inequality rigorously, the inequality can be tested to confirm its validity for various combinations of values for the variables a, b, c.

This can be done with the next algorithm:

Algorithm 2.2

```
tmp=0;
num_trials=10000000;

flag=0;

for k=1 to num_trials do
{
  a=rand(); b=rand(); c=rand();

  y1=(1-a^2*b)*(1-b^2*c)*(1-c^2*a);
  y2=(1-a^3)*(1-b^3)*(1-c^3);

  y=y1-y2;

  if(y<0)
  {
    flag=1;
    break;
  }

}
if (flag==0)
  print('The inequality has never been contradicted')
  else
     print('The inequality has been contradicted');
```

This algorithm has been implemented, and the results from 10 million simulation trials showed that the value of the flag had never changed to 1, which led to the output "The inequality has never been contradicted." The absence of a contradiction for 10 million random combinations of values for the variables a, b, and c is a strong indication that the conjectured inequality is probably true. As a result, an attempt for a rigorous proof of the inequality can be made. The described Monte-Carlo simulation Algorithm 2.1 was used to validate all algebraic inequalities presented in the book.

3 Obtaining New Physical Properties by Reverse Engineering of Algebraic Inequalities

The reverse engineering approach related to generating knowledge by interpreting an existing algebraic inequality has been outlined in Chapter 1. Attaching physical meaning to the variables in the abstract inequality combined with meaningful interpretation of the different parts of the inequality links the abstract inequality with the physical reality. As a result of the physical interpretation, often, the abstract inequality is a source of new knowledge or expresses a new physical property.

3.1 REVERSE ENGINEERING OF THE ARITHMETIC MEAN – HARMONIC MEAN INEQUALITY

Consider an example of the reverse engineering approach which starts with the abstract algebraic inequality

$$\left(x_1 + x_2 + \ldots + x_n \right) \ge n^2 \left(\frac{1}{1/x_1 + 1/x_2 + \ldots + 1/x_n} \right) \qquad (3.1)$$

which is valid for any set of n non-negative quantities x_i. Inequality (3.1) is equivalent to the inequality:

$$\frac{x_1 + x_2 + \ldots + x_n}{n} \ge \frac{n}{1/x_1 + 1/x_2 + \ldots + 1/x_n}$$

which is the Arithmetic-mean-Harmonic mean (AM-HM) inequality introduced in Chapter 2.

Inequality (3.1) can also be proved rigorously by reducing it to the standard Cauchy-Schwarz inequality, which states that for any two sequences of real numbers $\{a_1, a_2, \ldots, a_n\}$ and $\{b_1, b_2, \ldots, b_n\}$, the following inequality holds:

$$\left(a_1 b_1 + a_2 b_2 + \ldots + a_n b_n \right)^2 \le \left(a_1^2 + a_2^2 + \ldots + a_n^2 \right) \left(b_1^2 + b_2^2 + \ldots + b_n^2 \right) \qquad (3.2)$$

DOI: 10.1201/9781003517764-3

Because the quantities x_i in (3.1) are non-negative, the setting $a_i = \sqrt{x_i}$; $i = 1, n$ and $b_i = 1/\sqrt{x_i}$; $i = 1, n$ can be made. Next, substituting a_i and b_i in the Cauchy-Schwarz inequality (3.2) yields inequality (3.1).

To make inequality (3.1) relevant to a real system, appropriate physical meaning must be attached to the variables entering the inequality.

3.1.1 ELASTIC COMPONENTS AND RESISTORS CONNECTED IN SERIES AND PARALLEL

Suppose that the variables x_i in inequality (3.1) stand for the stiffness of an elastic element i. Now, the two sides of the inequality can be physically interpreted in the following way. The expression $x_1 + x_2 + \ldots + x_n$ on the left-hand side of the inequality (3.1) can be physically interpreted as the equivalent stiffness of n elastic elements connected in parallel (Figure 3.1a). The expression $\dfrac{1}{1/x_1 + 1/x_2 + \ldots + 1/x_n}$ on the right-hand side of the inequality can be physically interpreted as the equivalent stiffness of the same n elastic elements connected in series (Figure 3.1b). The inequality then establishes a connection between the equivalent stiffness of n elastic elements connected in parallel (the left-hand side) and the equivalent stiffness of the same n elastic elements connected in series (in the right-hand side).

According to the principle of consistency, the prediction from inequality (3.1) must be consistent with the properties of the series and parallel configurations of elastic elements. As a result, a new physical property can be inferred from inequality (3.1): the equivalent stiffness of n elastic elements connected in parallel is at least n^2 times

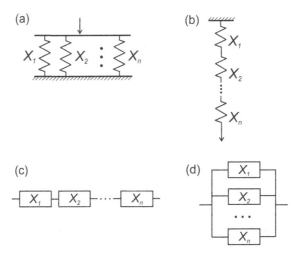

FIGURE 3.1 Reverse engineering of inequality 3.1 involving (a) elastic elements arranged in parallel (b) elastic elements arranged in series (c) resistors arranged in series and (d) resistors arranged in parallel.

larger than the equivalent stiffness of the same elastic elements connected in series, irrespective of the individual stiffness values of the elements (Todinov, 2019c).

The physical meaning created for the variables in inequality (3.1) is not unique and can be altered. Thus, a new physical meaning can be created if, for example, each x_i now stands for 'electrical resistance of element i'.

The expression $x_1 + x_2 + \ldots + x_n$ on the left side of inequality (3.1) can now be physically interpreted as the equivalent resistance of a system configuration including n elements connected in series (Figure 3.1c). On the right side of the inequality, the expression $\dfrac{1}{1/x_1 + 1/x_2 + \ldots + 1/x_n}$ can be physically interpreted as the equivalent resistance of another system configuration including the same n elements connected in parallel (Figure 3.1d). Inequality (3.1) now predicts another new physical property: the equivalent resistance of n elements connected in series is at least n^2 times larger than the equivalent resistance of the same elements connected in parallel, irrespective of the individual resistances.

This new physical property can be applied in testing the insulation of high-voltage equipment where the resistances of the insulating elements have extremely large values and are connected in series. Because of the very high resistances of the insulating elements, it is extremely difficult to measure directly not only the equivalent resistance $R_s = x_1 + x_2 + \ldots + x_n$ of n insulating elements connected in series but even the resistance x_i of a single insulating element. If, however, the insulating elements are connected in parallel (instead of series), it is much easier to measure the equivalent resistance $R_p = 1/(1/x_1 + 1/x_2 + \ldots + 1/x_n)$ of the parallel arrangement because, according to inequality (3.1), its magnitude is at least n^2 times smaller. After the measurement of the equivalent resistance R_p of a parallel arrangement, a conclusion about the equivalent resistance of the same elements connected in series can be made from the inequality $R_s \geq n^2 R_p$. The measured equivalent resistance R_s in series will be at least n^2 times larger than the measured equivalent resistance R_p of the same elements connected in parallel. This is exactly what is needed to assure the safety of the insulation assembly: the resistance of the series arrangement to be larger than a particular safe value.

It needs to be pointed out that for elements of equal resistances, the fact that the equivalent resistance in series is exactly n^2 times larger than the equivalent resistance of the resistors in parallel is a trivial result, known for a long period of time and used for high-resistance standards (Rozhdestvenskaya and Zhutovskii, 1968).

Indeed, for the equivalent resistance of n identical resistors connected in series, with resistances $x_1 = x_2 = \ldots = x_n = r$, the value nr is obtained. For the same n resistors connected in parallel, the value r/n is obtained. Clearly, the value nr is exactly n^2 times larger than the value r/n. However, the insight generated from the reverse engineering of inequality (3.1) leads to a much deeper result. *The relationship related to the equivalent resistances of a series and parallel connection holds for any possible values of the resistances of the individual elements* (Todinov, 2021a). The relationship given by inequality (3.1) does not require equal resistances. This new insight, derived from reverse engineering of inequality (3.1), eluded experts for decades.

3.1.2 Thermal Resistors and Electric Capacitors Connected in Series and Parallel

Identical reasoning applies to the problem related to the equivalent thermal resistance of n thermally conducting elements of different materials arranged in series (Figure 3.2a) and the equivalent thermal resistance of the same n elements arranged in parallel (Figure 3.2b). The equivalent thermal resistance R_s of n elements arranged in series, with thermal resistances $R_1, R_2, ..., R_n$, is given by (Tipler and Mosca, 2008)

$$R_s = R_1 + R_2 + ... + R_n \qquad (3.3)$$

while the equivalent thermal resistance R_p of the same elements arranged in parallel, is given by

$$R_p = \frac{1}{1/R_1 + 1/R_2 + ... + 1/R_n} \qquad (3.4)$$

If x_i in inequality (3.1) stands for the thermal resistance of the ith element, inequality (3.1) now predicts a different physical property: the equivalent thermal resistance of n elements in series is at least n^2 times larger than the equivalent thermal resistance of the same elements arranged in parallel, irrespective of the individual thermal resistances of the elements.

Now consider n electric capacitors and suppose that x_i stands for the capacitance of the ith capacitor. The expression $x_1 + x_2 + ... + x_n$ on the left-hand side of inequality (3.1) can be physically interpreted as equivalent capacitance of n capacitors connected in parallel. In the right-hand side of the inequality, the

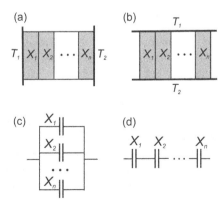

FIGURE 3.2 Interpretations of inequality 3.1 involving: (a) thermal elements arranged in series, (b) thermal elements arranged in parallel; (c) capacitors arranged in parallel and (d) capacitors arranged in series.

expression $\dfrac{1}{1/x_1 + 1/x_2 + \ldots + 1/x_n}$ can be physically interpreted as equivalent capacitance of the same n capacitors connected in series. Yet another physical property is predicted by inequality (3.1): the equivalent capacitance of n capacitors connected in parallel is at least n^2 times larger than the equivalent capacitance of the same capacitors connected in series, irrespective of the values of the individual capacitors (Todinov, 2021a).

The method of algebraic inequalities is domain-independent. It transcends mechanical engineering and can be used in other unrelated fields. In each application example, the left-hand side of inequality (3.1) was interpreted as an equivalent property of a system with elements connected in series or parallel, while the right-hand side of the inequality was interpreted as the same equivalent property characterizing a system built with the same elements but connected in parallel or series. The reverse engineering of inequality (3.1) establishes a fundamental link between the equivalent properties of the two different system configurations.

If multiple real experiments are performed connecting elastic elements, resistors, thermal elements, and capacitors in series and parallel, and the equivalent stiffness, electrical resistance, thermal resistance, and capacitance are measured, the observations will not contradict inequality (3.1). The statement expressed by inequality (3.1) and the behaviour of systems of springs, resistors, thermal elements, and capacitors arranged in series and parallel are consistent, and no contradiction will arise between the prediction of the algebraic inequality and the behaviour of the physical systems.

These examples illustrate how new knowledge is generated through reverse engineering of a correct algebraic inequality and the principle of consistency.

3.2   REVERSE ENGINEERING OF THE BERGSTRÖM INEQUALITY

3.2.1   Extensive Quantities and Additivity

Extensive quantities are typical examples of additive quantities and can be found in all areas of science and technology. Extensive quantities change with changing the size of their supporting objects/systems (DeVoe, 2012; Mannaerts, 2014). Extensivity always implies additivity. Additive quantities are, for example, mass, weight, amount of substance, number of particles, volume, distance, energy (kinetic energy, gravitational energy, electric energy, elastic energy, surface energy, internal energy), work, power, heat, force, momentum, electric charge, electric current, heat capacity, electric capacity, resistance (when the elements are in series), enthalpy and fluid flow. Additive quantities change with changing the size of the system.

Intensive quantities characterise the object/system locally and do not change with changing the size of the supporting system. Additivity is not present for intensive quantities. Additivity, for example, is not present for pressure or temperature. Consider, for example, a pressure vessel containing gas at a particular pressure and temperature. If a notional division of the pressure vessel is made, the temperature or pressure measured in the pressure vessel is not a sum of the temperature/pressure measured in the different parts of the vessel. Other properties where additivity is not present are 'density', 'concentration', 'hardness', 'velocity', 'acceleration', 'stress', 'surface tension', etc.

It is important to note that proportionality to mass is not a necessary condition for additivity. For a large group of extensive quantities (area, work, electric energy, elastic energy, displacement energy, surface energy, electric current, power, heat, electric charge), the proportionality to mass is absent (Mannaerts, 2014).

Depending on how the different elements composing a system are arranged, additivity may be present or absent. Thus, for resistances connected in series, additivity is present because the equivalent resistance of elements connected in series is a sum of the individual resistances. For resistances connected in parallel, additivity is absent.

Similarly, for voltage sources connected in series, additivity is present because the total voltage is a sum of the voltages of the individual sources. Additivity is also present for capacitors connected in parallel. For such an arrangement, the equivalent capacitance of the assembly is a sum of the individual capacitances of the capacitors. For capacitors connected in series, additivity is absent.

Normally, dividing two extensive quantities gives an intensive quantity (such as obtaining density by dividing mass to volume) and, as a result, additivity is absent but this is not always the case. Thus, dividing the extensive quantity 'volume' by the extensive quantity 'area' gives the extensive quantity length. Additionally, dividing the extensive quantity 'work' by the extensive quantity 'distance' over which the work has been done gives the extensive quantity 'force'.

3.2.2 REVERSE ENGINEERING OF THE BERGSTRÖM INEQUALITY RELATED TO ELECTRIC POWER OUTPUT

A special case of the general sub-additive function (2.38) is the Bergström inequality (Sedrakyan & Sedrakyan, 2010; Pop, 2009)

$$\frac{\left(a_1 + a_2 + \ldots + a_n\right)^2}{b_1 + b_2 + \ldots + b_n} \leq \frac{a_1^2}{b_1} + \frac{a_2^2}{b_2} + \ldots + \frac{a_n^2}{b_n} \tag{3.5}$$

valid for any sequences $a_1, a_2, \ldots, a_n$ and $b_1, b_2, \ldots, b_n$ of positive real numbers. The role of the sub-additive function $f(a, b)$ in inequality (2.38) is played by the function $f(a, b) \equiv a^2/b$ in inequality (3.5).

As shown in Section 2.2.4, Bergström inequality (3.5) is effectively a transformation of the well-known Cauchy-Schwarz inequality Cauchy-Schwarz inequality (Steele, 2004). Equality in (3.5) is attained only if $a_1/b_1 = a_2/b_2 = \ldots = a_n/b_n$. Inequality (3.5) can be reverse engineered in various domains of science and technology as long as the variables a_i, b_i and the terms a_i^2/b_i are additive quantities and have meaningful physical interpretation. To extract new knowledge from the reverse engineering of inequality (3.5), there is no need of any forward analysis. The condition for applying inequality (3.5) is the possibility to present an additive quantity p_i as a ratio of a square of an additive quantity a_i and another additive quantity b_i:

$$p_i = a_i^2 / b_i \qquad (3.6)$$

As an example, consider a case where factor a is 'voltage' from a source whose elements are arranged in series (an additive quantity) and factor b is 'equivalent resistance' of elements arranged in series (also an additive quantity). Suppose that the source of voltage V has been applied to n elements connected in series, with resistances $r_1, r_2, \ldots, r_n$ (Figure 3.3a).

Consider segmenting the source of voltage V into n smaller sources with voltages v_i such that $V = v_1 + \ldots + v_n$ (Figure 3.3b). Let $a_i = v_i$, $i = 1, \ldots, n$ be the smaller voltages applied to n separate elements with resistances r_i (Figure 3.3b). If $b_i = r_i$, condition (3.6) transforms into

$$p_i = v_i^2 / r_i \qquad (3.7)$$

which has a meaningful physical interpretation. In this equation, p_i is the power output [W] created by the ith voltage source, v_i is the voltage of the ith voltage source and r_i is the resistance of the ith element. In equation (3.7), all variables v_i, p_i and r_i are additive quantities.

Inequality (3.5) can be rewritten as

$$\frac{V^2}{r_1 + r_2 + \ldots + r_n} \leq \frac{v_1^2}{r_1} + \frac{v_2^2}{r_2} + \ldots + \frac{v_n^2}{r_n} \qquad (3.8)$$

The left-hand side of inequality (3.8) can now be physically interpreted as the power dissipated in the circuit from Figure 3.3a, while the right-hand side can be physically interpreted as the total power dissipated in the individual circuits from Figure 3.3b. The prediction from inequality (3.8) is that *the power output from a source with voltage V on elements connected in series is smaller than the total power output from the sources resulting from segmenting the*

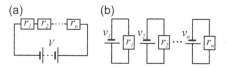

FIGURE 3.3   (a) A design option consisting of a single voltage source applied to elements connected in series; (b) a competing design option consisting of a voltage source V segmented into n smaller sources v_i applied to the individual elements.

original voltage source V and applying the segmented voltages to the individual elements (Todinov, 2021b).

This prediction holds irrespective of the individual resistances r_i of the elements and the voltage sources v_i into which the original voltage V has been segmented.

Note that the existence of asymmetry in the system is essential for increasing the power output through segmentation of the voltage source and the resistances. No increase in the power output is present if

$$v_1 / r_1 = v_2 / r_2 = \ldots = v_n / r_n = i \qquad (3.9)$$

is fulfilled. This means that combinations of resistances and sources resulting in the same current i in the segmented circuits do not bring an increase of the total power output.

To maximise the right-hand side of inequality (3.8), the squared voltages arranged in descending order $v_1^2 \geq v_2^2 \geq \ldots \geq v_n^2$ must be paired with the resistances arranged in ascending order ($r_1 \leq r_2 \leq \ldots \leq r_n$). For resistances arranged in ascending order, their reciprocals are arranged in descending order ($1/r_1 \geq 1/r_2 \geq \ldots \geq 1/r_n$) and according to the rearrangement inequality (see Section 2.1.5), the dot product $v_1^2 (1/r_1) + v_2^2 (1/r_2) + \ldots + v_n^2 (1/r_n)$ of two similarly ordered sequences is at its maximum.

The effect of voltage and resistance segmentation is significant. Thus, for resistances $r_1 = 10\Omega$, $r_2 = 15\Omega$, $r_3 = 25\Omega$, $r_4 = 50\Omega$ and voltage source of 16V, segmented into $v_1 = 6V$, $v_2 = 5V$, $v_3 = 3V$ and $v_4 = 2V$, the maximum possible power output is $P_{max} = v_1^2 / r_1 + v_2^2 / r_2 + v_3^2 / r_3 + v_4^2 / r_4 = 5.7W$. Any other permutation of voltages and resistances results in a smaller power output.

If the voltage source $V = 16V$ is not segmented, applying the voltage V to the four elements in series delivers output power $P = V^2/(r_1 + r_2 + r_3 + r_4) = 2.56W$, which is less than half the maximum power of $P_{max} = 5.7W$ delivered in the case of a voltage source segmentation.

The result from the reverse engineering of the algebraic inequality can be applied in electric circuits for heating. Despite that the circuits in Figure 3.3a and 3.3b seem to be different electric systems, they both can be viewed as alternative design options of a heating system. The circuit in Figure 3.3b dissipates more power (heat) for the same total number of voltage elements and the same set of resistive elements. It needs to be pointed out again that this advantage is not present if the segmented circuits are characterised by the same currents through the resistive elements.

3.2.3 REVERSE ENGINEERING OF THE BERGSTRÖM INEQUALITY RELATED TO STORED ELECTRIC ENERGY IN CAPACITORS

An alternative reverse engineering of inequality (3.5) can be created if the variable a stands for 'electric charge' and variable b stands for 'capacitance'.

In basic electronics (Floyd and Buchla, 2014), a well-known result is that the potential energy U stored by a charge Q in a capacitor with capacitance C is given by $U = Q^2/(2C)$. Note that the potential energy U, the charge Q and capacitance C of elements arranged in parallel are all additive quantities.

Consider segmenting the charge Q into n smaller charges q_i, $i = 1,...,n$, such that $Q = q_1 + ... + q_n$. Let $a_i = q_i$, $(i = 1, ..., n)$ in inequality (3.5) be the charges applied to n separate capacitors with capacitances C_i whose sum is equal to the capacitance of the original capacitor: $C = C_1 + C_2 + ... + C_n$. If $b_i = C_i$, inequality (3.5) can be rewritten as

$$\frac{Q^2}{2(C_1 + C_2 + ... + C_n)} \leq \frac{q_1^2}{2C_1} + \frac{q_2^2}{2C_2} + ... + \frac{q_n^2}{2C_n} \qquad (3.10)$$

Because $q_i^2 / (2C_i)$, q_i, C_i are all additive quantities, both sides of inequality (3.10) have a meaningful physical interpretation. The left-hand side of the inequality can be interpreted as the energy stored by a charge Q in a capacitor with capacitance C, while the right-hand side can be interpreted as the total energy stored in multiple capacitors with capacitances C_i, $(\Sigma_i C_i = C)$ by smaller charges q_i, $(\Sigma_i q_i = Q)$.

Inequality (3.5) now predicts that *the energy stored by a charge Q in a capacitor with capacitance C is smaller than the total energy stored in multiple capacitors with capacitances C_i, $\Sigma_i C_i = C$, by segmenting the charge Q into smaller charges and applying these to the individual capacitors* (Todinov, 2020c). This prediction holds irrespective of the individual capacitances C_i of the capacitors and the charges q_i into which the initial charge Q has been segmented.

Note that existence of asymmetry is absolutely essential for increasing the electrical energy stored by segmenting the initial charge Q. No increase in the stored electrical energy is present if $q_1/C_1 = q_2/C_2 = ... = q_n/C_n = v$. This means that capacitors loaded to the same potential difference $v = q_i/C_i$ on the plates, do not yield an increase of the amount of stored total electrical energy.

To maximise the right-hand side of inequality (3.10), the squared segmented charges arranged in descending order $q_1^2 \geq q_2^2 \geq ... \geq q_n^2$ must be paired with the capacitances arranged in ascending order $(C_1 \leq C_2 \leq ... \leq C_n)$. For capacitances arranged in ascending order, the reciprocals $1/(2C_1) \geq 1/(2C_2) \geq ... \geq 1/(2C_n)$ are arranged in descending order and, according to the rearrangement inequality, the 'dot product' $q_1^2/(2C_1) + q_2^2/(2C_2) + ... + q_n^2/(2C_n)$ of two similarly ordered sequences is at its maximum (see Section (2.1.5).

The segmentation of additive quantities through the algebraic inequality (3.5) can be used to develop systems and processes with superior performance and the algebraic inequality is applicable in any area of science and technology, as long as the additive quantities comply with the simple condition (3.6).

It is important to note that inequality (3.5) can be applied to each of the individual terms a_i^2/b_i on the left-hand side which in turn can be segmented. The result from this recursive segmentation is further multiplication of the effect from the segmentation.

Interestingly, no such properties have been reported in modern comprehensive texts in the field of electronics (Floyd and Buchla, 2014; Horowitz and Hill, 2015). This demonstrates that the meaningful interpretation of the abstract inequality (3.5) helped discover an overlooked property in a mature field like electronics.

It might seem that segmenting the applied voltage would maximize the energy stored in a set of capacitors, much like how segmenting the applied voltage maximizes the output power in a set of resistors. However, this conjecture is false.

Indeed, the electrical energy U stored in a set of capacitors with capacitances $C_1, C_2, ..., C_n$, arranged in parallel (Figure 3.4) by an applied voltage of magnitude V is given by (Floyd and Buchla, 2014)

$$U_1 = \frac{1}{2} C V^2 \tag{3.11}$$

where $C = C_1 + C_2 + ... + C_n$.

Segmenting the voltage source V and applying the smaller voltage sources v_i ($v_1 + v_2 + ... + v_n = V$) to the individual capacitors result in total stored energy U_2 given by

$$U_2 = \frac{1}{2} C_1 v_1^2 + \frac{1}{2} C_2 v_2^2 + ... + \frac{1}{2} C_n v_n^2 \tag{3.12}$$

By a direct expansion, it can be seen that

$$\frac{1}{2} C_1 v_1^2 + \frac{1}{2} C_2 v_2^2 + ... + \frac{1}{2} C_n v_n^2 \le \frac{1}{2} \left(C_1 + C_2 + ... C_n \right) \left(v_1 + v_2 + ... + v_n \right)^2 \tag{3.13}$$

Therefore $U_2 \le U_1$ which means that the segmentation of a voltage source over a number of capacitors is associated with a decrease of the total stored energy. In the case of voltage sources applied to capacitors, the aggregation of the voltage sources and the capacitors leads to an increase of the total stored energy.

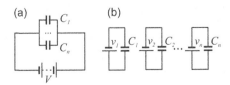

FIGURE 3.4 (a) A single voltage source applied to capacitors connected in parallel; (b) A voltage source V segmented into n smaller sources v_i applied to the individual capacitors.

3.2.4 Reverse Engineering of the Bergström Inequality Related to Accumulated Elastic Strain Energy

3.2.4.1 Increasing the Accumulated Elastic Strain Energy for Components Loaded in Tension

An alternative reverse engineering of the sub-additive inequality (3.5) can also be made if variable a, for example, stands for the additive quantity 'force' and variable b stands for the additive quantity 'area'.

It is a well-known result from mechanics of materials (Hearn, 1985; Beer et al., 2002) that the accumulated strain energy U of a linearly elastic bar with length L and cross-sectional area A is given by the equation:

$$U = \frac{P^2 L}{2EA} \tag{3.14}$$

where E [Pa] denotes the Young's modulus of the material and P [N] is the magnitude of the loading force. The strain energy U [J] is an additive quantity. Equation (3.14) can be written as

$$U = \frac{P^2}{2k} \tag{3.15}$$

where $k = EA/L$ [N/m] is the stiffness of the bar.

Consider the two system configurations in Figure 3.5. In the system configuration from Figure 3.5a, a single force P acts on a single bar with cross-sectional area A. In the design configuration from Figure 3.5b, the original bar has been segmented into n individual bars with smaller cross sections A_1, A_2, ..., A_n, the sum of which is equal to the cross-section A of the original bar ($A = A_1 + \ldots + A_n$).

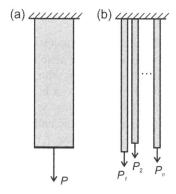

FIGURE 3.5   (a) A force P applied to a single bar; (b) Segmented forces P_i, applied to the individual bars into which the original bar has been segmented (the individual bars are loaded independently and do not necessarily have the same displacements and stresses).

Similarly, the force P has also been segmented into n forces $P_1, P_2, \ldots, P_n$, the sum of which is equal to the initial loading force P ($P = P_1 + P_2 + \ldots + P_n$). The smaller forces P_i have been applied to the individual bars independently, and the individual bars do not necessarily have the same displacements and stresses (Figure 3.5b). The support is rigid and does not deform due to the segmented forces. A question of interest is which design configuration in Figure 3.5 is capable of accumulating more elastic strain energy.

Let $a_i = P_i$, $i = 1, \ldots, n$ be the forces applied to the centroids of the n separate bars whose stiffness values are $k_i = E_i A_i / L_i$ (Figure 3.5b). Let also $b_i = k_i$, $i = 1, \ldots, n$.

The system configurations in Figure 3.5a and 3.5b are not equivalent mechanically and achieving equivalence is not the purpose of the load segmentation. The purpose of the load segmentation and the bar segmentation is to increase the elastic strain energy stored in the system. This is why, there is no requirement about equivalence of moments. The only requirement is the sum of the magnitudes of the segmented forces P_i to be equal to the magnitude of the original loading force P and the sum of the cross-sectional areas A_i of the segmented bars to be equal to the cross-sectional area A of the original bar.

Because, the supports in Figure 3.5a and b are rigid, no deformations exist in the supports and the elastic strain energy of the system of bars is a sum of the elastic strain energies $P_i / (2k_i)$ of the individual bars only. As a result, the total elastic strain energy U_{seg} of the multiple bars in Figure 3.5b is: $U_{seg} = P_1^2 / (2k_1) + \ldots + P_n^2 / (2k_n)$ while the strain energy U_0 of the single solid bar in Figure 3.5a is: $U_0 = P^2 / [2(k_1 + k_2 + \ldots + k_n)]$. Calculating the total elastic strain energy of the system of multiple bars in Figure 3.5b has been done by applying Equation (3.15) n times, where n is the number of segmented bars.

The elastic strain energies of the two system configurations can now be compared through inequality (3.5). If $b_i = k_i$, inequality (3.5) can be rewritten as

$$\frac{P^2}{2\left(k_1 + k_2 + \ldots + k_n\right)} \leq \frac{P_1^2}{2k_1} + \ldots + \frac{P_n^2}{2k_n} \tag{3.16}$$

The left-hand side of inequality (3.16) corresponds to the system configuration in Figure 3.5a, and can be physically interpreted as accumulated elastic strain energy due to a loading force P acting on a single bar with stiffness $k = k_1 + k_2 + \ldots + k_n$.

The right-hand side of inequality (3.16) can be physically interpreted as the accumulated elastic strain energy resulting from segmenting the original force P into smaller forces P_i and applying these to the individual bars with stiffnesses k_i into which the original bar has been segmented.

Inequality (3.16) now predicts that *the accumulated elastic strain energy from a loading force acting on a single bar is smaller than the accumulated elastic strain energy, resulting from segmenting the loading force into smaller forces applied to the individual bars into which the original bar has been segmented* (Todinov, 2022c).

It needs to be pointed out that not every load segmentation of the original bar leads to an increase of the accumulated elastic strain energy. If all loaded bars have the same displacement, no increase in the accumulated elastic strain energy is present. Indeed, in inequality (3.16), the value $\delta_i = P_i/k_i$ is the displacement of the ith bar under load P_i. If all bars displacements are equal ($\delta_1 = \delta_2 = \ldots = \delta_n = \delta$), it is easy to verify that equality is attained in (3.16). Indeed, the right-hand side of

inequality (3.16) gives $\dfrac{1}{2}\left(\dfrac{P_1^2}{k_1} + \ldots + \dfrac{P_n^2}{k_n}\right) = (1/2)\delta\left(P_1 + P_2 + \ldots + P_n\right) = (1/2)\delta \times P.$

Since $P = P_1 + \ldots + P_n = k_1\delta + \ldots + k_n\delta = \delta(k_1 + \ldots + k_n)$, it follows that $P/(k_1 + \ldots + k_n) = \delta$ and for the left-hand side of inequality (3.16) $P^2/[2(k_1 + \ldots + k_n)] = (1/2)\delta \times P$ holds. Hence, equality is, indeed, attained in inequality (3.16). Note that the loading in Figure 3.5a and 3.5b differs significantly. In Figure 3.5b, the bar and the force have been split into smaller bars and forces and each individual bar has been loaded independently. As a result, the stresses and displacements across the segmented bars are no longer uniform.

Inequality (3.16) holds irrespective of the magnitude of the separate forces into which the initial force P has been segmented and irrespective of the cross sections of the individual bars. The inequality provides a method of increasing the capacity for absorbing elastic strain energy upon dynamic loading.

This is an unexpected result. To illustrate its validity, consider an example of a steel bar with cross-section 10mm x 20mm and length 1m, loaded with a force of 50 kN. The Young's modulus of the material is 210 GPa and the yield strength of the material is 650 MPa.

According to equation (3.15), the accumulated elastic strain energy in the bar with stiffness $k = EA/L$, loaded with a force $P_0 = 50$ kN is given by

$$U_0 = \frac{P_0^2}{2k} = \frac{\left(50 \times 10^3\right)^2}{2 \times \left[\left(210 \times 10^9 \times 20 \times 10 \times 10^{-6}\right)/1\right]} = 29.76J$$

Now, suppose that the bar has been segmented into two bars with cross-sections 10mm × 10mm and length 1m, and the force P_0 has been segmented into two *unequal* forces with magnitudes 40 kN, and 10 kN, applied to the individual bars. In this case, the accumulated elastic strain energy in the bars is

$$U_1 = \frac{P_1^2}{2k_1} + \frac{P_2^2}{2k_2} = \frac{\left(40 \times 10^3\right)^2}{2 \times \left[210 \times 10^9 \times 10 \times 10 \times 10^{-6}/1\right]}$$

$$+ \frac{\left(10 \times 10^3\right)^2}{2 \times \left[210 \times 10^9 \times 10 \times 10 \times 10^{-6}/1\right]} = 38.1 + 2.38 = 40.48J$$

This is about 36% larger than the elastic strain energy of 29.76 J accumulated in the single bar.

Increasing the capability of accumulating strain energy is important not only in cases of preventing failure during dynamic loading but also in cases where more elastic strain energy needs to be stored.

If the bar had been segmented into two bars with cross sections of 8 mm × 10 mm and 12 mm × 10 mm, and the force had been segmented into two forces with magnitudes of 20 kN and 30 kN, the accumulated elastic strain energy characterizing the segmented bar would be equal to the strain energy characterizing the original bar:

$$U_1 = \frac{P_1^2}{2k_1} + \frac{P_2^2}{2k_2} = \frac{\left(20 \times 10^3\right)^2}{2 \times \left[210 \times 10^9 \times 8 \times 10 \times 10^{-6}/1\right]}$$

$$+ \frac{\left(30 \times 10^3\right)^2}{2 \times \left[210 \times 10^9 \times 12 \times 10 \times 10^{-6}/1\right]} = 29.76J$$

This is because the displacements of the segmented bars are equal (the ratios $P_1/k_1 = P_2/k_2 = \delta$, are the same) and, according to the basic properties of inequality (3.5), in this case, equality is attained.

If the bar had been segmented into two bars with cross sections 10mm × 10mm and the force had been segmented into two equal forces with magnitude 25kN, again, the accumulated elastic strain energy in the bar would be the same:

$$U_1 = \frac{P_1^2}{2k_1} + \frac{P_2^2}{2k_2} = 2 \times \frac{\left(25 \times 10^3\right)^2}{2 \times \left[210 \times 10^9 \times 10 \times 10 \times 10^{-6}/1\right]} = 29.76J$$

because the displacements of the bars $P_1/k_1 = P_2/k_2 = \delta$ are the same.

In order to obtain any advantage, according to inequality (3.5), *asymmetry must be present in the system* so *that the displacements of the bars are not equal*: $\delta_1 = P_1/k_1 \neq \delta_2 = P_2/k_2$. Existence of asymmetry is absolutely essential for increasing the accumulated elastic strain energy through segmentation. Segmented bars experiencing the same displacement do not yield an increase of the amount of stored elastic strain energy.

The asymmetry requirement to increase the elastic energy accumulation is rather counterintuitive and makes this result difficult to obtain by alternative means bypassing inequality (3.5).

Consider the sequence $\{P_1^2, P_2^2, ..., P_n^2\}$ and the sequence $\{1/(2k_1), 1/(2k_2), ..., 1/(2k_n)\}$. The dot product of these sequences constitutes the right-hand side of inequality (3.16). According to the rearrangement inequality, the dot product of two sequences is maximized if they are similarly ordered, for example, if $P_1^2 \leq P_2^2 \leq ... \leq P_n^2$ and $1/(2k_1) \leq 1/(2k_2) \leq, ..., \leq 1/(2k_n)$. For the first sequence, it can be shown that if $P_1^2 \leq P_2^2 \leq ... \leq P_n^2$ then $P_1 \leq P_2 \leq ... \leq P_n$. Indeed, from the basic properties of inequalities, for positive P_i and P_j from $P_i^2 \leq P_j^2$ it follows that $P_i \leq P_j$.

For the second sequence, it is easy to see that $1/k_1 \leq 1/k_2 \leq, \ldots, \leq 1/k_n$ only if $k_1 \geq k_2 \geq \ldots \geq k_n$. For other permutations of the sequences, for example $\{P_{k1}, P_{k2}, \ldots, P_{kn}\}$ and $\{k_{s1}, k_{s2}, \ldots, k_{sn}\}$, the next inequality is fulfilled:

$$\frac{P_1^2}{2k_1} + \ldots + \frac{P_n^2}{2k_n} \geq \frac{P_{k1}^2}{2k_{s1}} + \ldots + \frac{P_{kn}^2}{2k_{sn}} \qquad (3.17)$$

As a result, the right-hand side of inequality (3.16) is maximised if the ordered in ascending order segments $P_1 \leq P_2 \leq \ldots \leq P_n$ of the loading force are paired with the ordered in descending order stiffness values: $k_1 \geq k_2 \geq \ldots \geq k_n$.

3.2.4.2 Increasing the Accumulated Elastic Strain Energy for Components Loaded in Bending

It is a well-known result from mechanics of materials that the accumulated elastic strain energy U in a cantilever elastic beam (Figure 3.6a) with rectangular cross section $b \times h$ and length L is given by the equation:

$$U = \frac{1}{2} P \times f \qquad (3.18)$$

where P denotes the loading force and f is the deflection of the beam at the point of application of the concentrated force P.

From mechanics of materials (Gere and Timoshenko, 1999; Hearn, 1985; Beer et al., 2002), for a cantilever beam with rectangular cross-section, the deflection f can be determined from

$$f = \frac{PL^3}{3EI} \qquad (3.19)$$

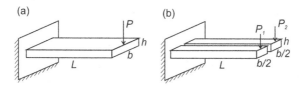

FIGURE 3.6 (a) Cantilever beam loaded with a single concentrated force P; (b) segmented cantilever beam loaded with two forces P_1 and P_2 whose sum is equal to the original force P. (The segmented beams are loaded independently and do not necessarily have the same displacements and stresses).

where E is the Young's modulus of the material and $I = bh^3/12$ is the second moment of area of the beam. Substituting I in Equation (3.19) gives

$$f = \frac{4PL^3}{Ebh^3} \qquad (3.20)$$

According to equations (3.18) and (3.20), the accumulated elastic strain energy U in the cantilever elastic beam is then given by

$$U = \frac{2P^2L^3}{Ebh^3} \qquad (3.21)$$

By introducing the variable $k = \dfrac{Ebh^3}{4L^3}$, standing for the *flexural stiffness* of the cantilever beam, Equation (3.21) for the accumulated elastic strain energy can be written as

$$U = \frac{P^2}{2k} \qquad (3.22)$$

Suppose that the load P and the cantilever beam have been segmented into n loads $P_1, P_2, \ldots, P_n \left(\sum_{i=1}^{n} P_i = P \right)$ and n beams with the same thickness h and smaller widths $b_1, b_2, \ldots, b_n \left(\sum_{i=1}^{n} b_i = b \right)$. For $n = 2$, such segmentation is shown in Figure 3.6b.

In this case, the flexural stiffness k of the unsegmented beam is equal to the sum of the flexural stiffnesses characterising the smaller cantilever beams: $k = k_1 + k_2 + \ldots + k_n$, where $k_i = \dfrac{Eb_i h^3}{4L^3}$. Indeed,

$$k_1 + k_2 + \ldots + k_n = \frac{Eh^3}{4L^3} \sum_{i=1}^{n} b_i = \frac{Eh^3 b}{4L^3} = k$$

According to inequality (3.5):

$$\frac{P_1^2}{2k_1} + \frac{P_2^2}{2k_2} + \ldots + \frac{P_n^2}{2k_n} \geq \frac{\left(P_1 + P_2 + \ldots + P_n \right)^2}{2 \left(k_1 + k_2 + \ldots + k_n \right)} \qquad (3.23)$$

The right-hand side of inequality (3.23) can be physically interpreted as the accumulated elastic strain energy due to a load P acting on a cantilever beam with flexural stiffness $k = k_1 + k_2 + \ldots + k_n$. This design option is shown in Figure 3.6a. The left-hand side of inequality (3.23) can be physically interpreted as total accumulated strain energy from loads $P_i \left(\sum_i P_i = P \right)$, resulting from segmenting the original load P into smaller loads and applying the smaller loads to the smaller individual cantilever beams with flexural stiffnesses k_i. For $n = 2$, this design option is shown in Figure 3.6b.

Inequality (3.23) predicts that the accumulated elastic strain energy due to a load acting on an unsegmented cantilever beam is smaller than the total accumulated elastic strain energy from the smaller loads from segmenting the original load and applying the segmented loads on the cantilever beam segments (Todinov, 2022c).

Similar to the previous example related to bars, not every segmentation of the beam and the load leads to an increase of the accumulated elastic strain energy. If the cantilever beam segments have the same displacement under their loads, no increase in the accumulated strain energy will be present. Indeed, in inequality (3.23), the value $f_i = P_i/k_i$ is the displacement of the ith cantilevered beam under load P_i. If all beam displacements are equal $(f_1 = f_2 = \ldots = f_n = f)$, equality is attained in (3.23). Indeed, the left-hand side of (3.23) gives $\dfrac{P_1^2}{2k_1} + \ldots + \dfrac{P_n^2}{2k_n} = (f/2)(P_1 + P_2 + \ldots + P_n) = (f/2) \times P$. Since, $P_1 = k_1 f$, $P_2 = k_2 f, \ldots, P_n = k_n f$ and $P_1 + \ldots + P_n = (k_1 + \ldots + k_n)f$, it follows that $P/(k_1 + \ldots + k_n) = f$ and the right-hand side of inequality (3.23) gives $P^2/[2(k_1 + \ldots + k_n)] = (f/2) \times P$. Hence, equality is indeed attained in (3.23).

Again, the requirement for asymmetry to achieve the effect of increased elastic strain energy is rather counterintuitive and makes this result difficult to obtain by alternative means, bypassing the use of the inequality (3.5).

The reverse engineering of the abstract inequality (3.5) helped find overlooked properties in the mature field of mechanical engineering. No such result has been reported in modern comprehensive textbooks in the area of mechanical engineering and stress analysis (French, 1999; Samuel and Weir, 1999; Mott et al., 2018; Pahl et al., 2007; Gullo and Dixon, 2018; Thompson, 1999; Gere and Timoshenko, 1999; Budynas, 1999; Beer et al., 2002; Childs, 2014; Budynas and Nisbett, 2015; Maier et al. 2022; Ugural, 2022).

New knowledge is readily generated by reverse engineering of inequalities based on sub-additive functions, as long as the variables in the sub-additive functions and the separate terms represent additive quantities. The obtained physical insights can then be used to optimize various systems and processes in any area of science and technology.

4 Light-Weight Designs and Improving the Load-Bearing Capacity of Structures by Reverse Engineering of Algebraic Inequalities

4.1 AN OVERLOOKED APPROACH FOR CREATING LIGHTWEIGHT DESIGNS

The sub-additive and super-additive algebraic inequalities based on power laws, the properties of which have been discussed in Section 2.2.5, are excellent candidates for reverse engineering. Power laws are ubiquitous in describing physical phenomena; therefore, sub-additive and super-additive algebraic inequalities based on power laws are excellent candidates for reverse engineering.

Lightweight components can be obtained through design, manufacturing, and lightweight materials. A major trend in lightweight parts obtained through design is the simulation-driven design technique known as 'topology optimization' (Walton and Moztarzadeh, 2017; Long et al., 2020). In addition, lightweight components can also be obtained by using advanced manufacturing technologies such as advanced metal forming (Rosenthal et al., 2020) and additive manufacturing (Yang et al., 2016; Mandolini et al., 2022).

There is however, another powerful yet overlooked method for improving the load-bearing capacity of structures and producing lightweight structures. This method will be referred to as the method of aggregation and was discovered through reverse engineering of algebraic inequalities (Todinov, 2022a, 2024a). Accordingly, this chapter demonstrates huge material savings and increase of the load-carrying capacity of structures obtained through the method of aggregation. As a result, the method of aggregation is directly related to structural design.

For multiple, uniformly loaded identical elements we can distinguish two alternative arrangements. A non-aggregated structure, composed of n uniformly loaded elements and an aggregated structure composed of a smaller number m of uniformly loaded elements with larger cross-sections. As shall be demonstrated later, aggregating uniformly loaded elements into fewer uniformly loaded elements

 DOI: 10.1201/9781003517764-4

with larger, geometrically similar cross-sections, opens a huge opportunity for material saving. Despite the widespread use of multiple load-carrying elements in engineering and construction, there is a surprising deficiency in the analysis of non-aggregated and aggregated structures. Notably, standard textbooks on stress analysis and machine design (Budynas and Nisbett, 2015; Collins, 2003; Budynas, 1999; Hearn, 1985; Gere and Timoshenko, 1999; Beer et al., 2002), lack a key discussion. This is related to the comparison of total volume of material needed to support a load of specified magnitude for alternative structures based on varying numbers of load-carrying elements. Similarly, no analysis is present related to the load-carrying capacity of non-aggregated and aggregated structure built with the same volume of material. Surprisingly, this critical discussion is absent not only in textbooks dedicated to structural engineering (Hibbeler 2019; Podder and Chatterjee 2022), but also in the structural reliability literature and papers focusing on the optimization of loaded beams (Wang 2020; Kiureghian 2022).

4.2 REVERSE ENGINEERING OF SUB-ADDITIVE AND SUPER-ADDITIVE INEQUALITIES BASED ON A CONCAVE AND CONVEX POWER LAW DEPENDENCE

Consider the sub-additive inequality (Todinov, 2024a)

$$ax_1^p + ax_2^p + \ldots + ax_m^p < ay_1^p + ay_2^p + \ldots + ay_n^p \qquad (4.1)$$

where $m < n$, $p < 1$, $x_i > 0$, $y_i > 0$, $a > 0$ and $\sum_{i=1}^{m} x_i = \sum_{i=1}^{n} y_i$. In general, inequality (4.1) holds under the following sufficient majorising conditions:

$$\begin{aligned}
x_1 \geq y_1; \ x_1 \geq y_2; \ldots; x_1 \geq y_n \\
x_2 \geq y_1; \ x_2 \geq y_2; \ldots; x_2 \geq y_n \\
x_m \geq y_1; \ x_m \geq y_2; \ldots; x_m \geq y_n
\end{aligned} \qquad (4.2)$$

The majorising conditions (4.2) effectively state that any x_i in the left-hand side of inequality (4.1) majorises each y_i in the right-hand side of the inequality. (In general, x_i and y_i are not necessarily equal). The proof of inequality (4.1) under these sufficient conditions has been given in Section 2.2.5.1.

For equal x_i and y_i ($x_1 = x_2 = \ldots = x_m = z/m$ and $y_1 = y_2 = \ldots = y_n = z/n$), the sufficient majorising conditions (4.2) are automatically satisfied and inequality (4.1) holds true.

For $p > 1$, $m < n$, $x_i > 0$, $y_i > 0$, $a > 0$, $\sum_{i=1}^{m} x_i = \sum_{i=1}^{n} y_i$ and under the same sufficient conditions (4.2) specified for inequality (4.1), the super-additive inequality

$$ax_1^p + ax_2^p + \ldots + ax_m^p > ay_1^p + ay_2^p + \ldots + ay_n^p \qquad (4.3)$$

holds true.

The primary advantage of inequalities (4.1) and (4.3) lies in their simplicity and ease of physical interpretation which renders them particularly suitable for reverse engineering. In inequalities (4.1) and (4.3), x and y are additive quantities while 'a' and 'p' are constants. For a meaningful physical interpretation of inequalities (4.1) and (4.3), each individual term ax_i^p and ay_i^p within the inequalities must also be an additive quantity.

Inequality (4.1) can be reverse engineered if the variables x_i and y_i are interpreted as load magnitudes for an element loaded in bending. The terms ax_i^p and ay_i^p in (4.1) can be physically interpreted as the minimal volume of material necessary to support bending loads of magnitudes x_i and y_i, where $P = \sum_{i=1}^{m} x_i = \sum_{i=1}^{n} y_i$ is the total bending load carried by each structure, incorporating m and n load-carrying elements, correspondingly. Consequently, the left-hand side of inequality (4.1) represents the minimal volume of a structure built with m elements loaded in bending, necessary to carry a total load $P = \sum_{i=1}^{m} x_i$. The right-hand side of the inequality represents the minimal volume of an alternative structure built with n elements loaded in bending, necessary to carry the same total load $P = \sum_{i=1}^{n} y_i$. As a result, the physical interpretation of inequality (4.1) provides a mechanism for comparing the volumes of material necessary to carry a specified load for competing structures loaded in bending and selecting the structure characterized by the smaller volume of material.

Just as the reverse engineering of the sub-additive inequality (4.1) compares the minimum volumes of material necessary to carry the same total bending load, the reverse engineering of the super-additive inequality (4.3) compares the load-bearing capacities at the same total volume of material used for building the competing structures. In this case, the variables x_i and y_i are physically interpreted as 'volumes of the elements loaded in bending, building the structures'. The terms ax_i^p and ay_i^p in (4.3) are physically interpreted as 'the bending loads supported by the individual elements of the alternative structures' whose volumes are x_i ($i = 1,...,$ m) and y_i ($i = 1,...,n$), respectively. In this case, the volume of material used for each of the competing structures is the same: $V = \sum_{i=1}^{m} x_i = \sum_{i=1}^{n} y_i$. As a result, the left-hand side of the super-additive inequality (4.3) represents the total bending load P_m carried by a structure built with m elements loaded in bending: $P_m = \sum_{i=1}^{m} ax_i$. The right-hand side of inequality (4.3) represents the total bending load P_n carried by an alternative structure built with the same total volume of material, that includes n elements loaded in bending: $P_n = \sum_{i=1}^{n} ay_i$. The physical interpretation of inequality (4.3) provides a mechanism for comparing the load-bearing capacities of competing structures loaded in bending, at the same volume of material used for building the structures. In summary, the reverse engineering of inequalities (4.1) and (4.3) provides the theoretical basis for the method of aggregation in developing lightweight designs and increasing the load-bearing capacity of structures under bending loads (Todinov, 2024a). This new insight uncovered through reverse engineering of algebraic inequalities eluded stress analysis experts for decades.

Although the inequalities (4.1) and (4.3) are applicable to non-uniform loads and cross-sections, this chapter illustrates the power of the aggregation method by focusing on uniformly distributed loads and uniform cross-sections of the load-bearing elements. This choice is deliberate for several reasons. Firstly, the uniform distribution automatically fulfils the majorising conditions (4.2), sufficient for the validity of inequalities (4.1) and (4.3). Secondly, it encompasses an important special case with practical importance. Thirdly, it makes it possible to easily quantify the effects of applying aggregation. Lastly, the uniform setup allows for a straightforward treatment, facilitating better understanding of the aggregation method.

4.3 STRUCTURES BUILT ON CANTILEVERED BEAMS LOADED IN BENDING

Consider two alternative structures, each including m and n identical load-bearing beams, correspondingly, where $m < n$ (Figure 4.1a and b). The length L of the beams is the same for both structures. The aggregated structure (Figure 4.1a) differs from the non-aggregated structure (Figure 4.1b) in that it contains fewer cantilevered beams with larger, geometrically similar cross-sections. For the sake of simplicity of the derivations and presentation, variations in material properties and geometry are not considered.

4.3.1 MINIMUM VOLUME OF MATERIAL NEEDED FOR SUPPORTING A TOTAL BENDING LOAD OF A SPECIFIED MAGNITUDE

Suppose that the critical tensile stress permitted by the material of both structures in Figure 4.1 is σ_{cr}. Consider a uniform distribution of the bending load P over the load-bearing beams. In this case, the bending loads per beam for the aggregated and non-aggregated structure are P/m and P/n, respectively. We will calculate the minimum volume of material for the aggregated structure (Figure 4.1a) and the non-aggregated structure (Figure 4.1b), needed to support a total bending load of given magnitude P.

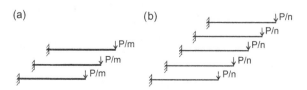

FIGURE 4.1   (a) Aggregated and (b) non-aggregated structure based on cantilever beams.

For a cantilevered beam with radius r, subjected to a bending moment M, the absolute value of the maximum stress σ is given by the familiar expression (Gere and Timoshenko, 1999; Beer et al., 2002):

$$\sigma = \left(M / I \right) r \qquad (4.4)$$

where I is the second moment of area of the circular cross section. Since for a circular section, $I = \pi r^4/4$, expression (4.4), solved with respect to the radius r gives:

$$r = \left[4M / \left(\pi\sigma \right) \right]^{1/3}$$

The loading moment at the fixed end of a cantilevered beam from the aggregated structure in Figure 4.1a is $(P/m) \times L$ and for the non-aggregated structure in Figure 4.1b, the loading moment is $(P/n) \times L$. Assume that the same permissible critical tensile stress σ_{cr} is developed in the beams from both structures. Then, for the radius r_1 of the load-bearing beams from the aggregated structure the expression:

$$r_1 = \left[4PL / \left(m\pi\sigma_{cr} \right) \right]^{1/3}$$

is obtained while for the radius r_2 of the load-bearing beams of the non-aggregated structure, the expression:

$$r_2 = \left[4PL / \left(n\pi\sigma_{cr} \right) \right]^{1/3}$$

is obtained. The volume of a single beam from the aggregated structure is therefore given by:

$$v_1 = \pi r_1^2 L = \pi L \left[4PL / \left(m\pi\sigma_{cr} \right) \right]^{2/3} = \pi L \left[4L / \left(\pi\sigma_{cr} \right) \right]^{2/3} \times \left[P / m \right]^{2/3} \quad (4.5)$$

while the volume of a single beam from the non-aggregated structure is given by

$$v_2 = \pi r_2^2 L = \pi L \left[4PL / \left(n\pi\sigma_{cr} \right) \right]^{2/3} = \pi L \left[4L / \left(\pi\sigma_{cr} \right) \right]^{2/3} \times \left(P / n \right)^{2/3}$$

If we denote $x_i = P/m$, $y_i = P/n$, $\left(\sum_{i=1}^{m} x_i = \sum_{i=1}^{n} y_i = P \right)$ and $a = \pi L[4L/(\pi\sigma_{cr})]^{2/3}$, the majorising conditions (4.2) for inequality (4.1) will be fulfilled and it becomes:

$$ax_1^{2/3} + \ldots + ax_m^{2/3} < ay_1^{2/3} + \ldots + ay_n^{2/3} \qquad (4.6)$$

The physical interpretation of inequality (4.6) then yields the following. The total volume $V_1 = ax_1^{2/3} + ax_2^{2/3} + \ldots + ax_m^{2/3}$ of the beams in the aggregated structure needed to support the total load P is smaller than the total volume $V_2 = ay_1^{2/3} + ay_2^{2/3} + \ldots + ay_n^{2/3}$ of the beams in the non-aggregated structure needed to support the same total load P. Considering that the left-hand side of (4.6) is equal to $V_1 = ma(P/m)^{2/3}$ and the right-hand side of (4.6) is equal to $V_2 = na(P/n)^{2/3}$, the ratio of the two volumes is given by

$$V_2 / V_1 = \left(n / m \right)^{1/3} \qquad (4.7)$$

The material saving given by equation (4.7) is very big. According to equation (4.7), for $m = 1$, the material saving factor s_f is given by the expression:

$$s_f = n^{1/3} \qquad (4.8)$$

where n is the number of load-carrying beams in the non-aggregated structure. Evaluating expression (4.8), for $n = 2, 3, 4, 5, 6, \ldots$, yields the material saving factors $s_f = 1.26, 1.44, 1.587, 1.71, 1.82\ldots$. These indicate that applying the method of aggregation yields huge material savings (Todinov, 2024a).

For cantilever beams, the material saving factors achieved by the method of aggregation have been summarised in Figure 4.2.

4.3.2 COMPARISON AND VERIFICATION

For the purposes of the comparison and verification of these theoretical results, the following three variants of aggregation were analysed. First, two cylindrical cantilevered beams supporting a specified total bending load P were considered, aggregated into a single cylindrical cantilevered beam with a $2^{1/3}$ times smaller volume, supporting the same total bending load, P (Figure 4.3a). Next, three beams supporting a specified total load P were considered, aggregated into a single cantilevered beam with a $3^{1/3}$ times smaller volume, supporting the same total bending load P (Figure 4.3b). Finally, four beams supporting a specified total load P were considered, aggregated into a single beam with a $4^{1/3}$ times smaller volume carrying the same total bending load P (Figure 4.3c).

In the first variant of aggregation, the non-aggregated structure is composed of two cylindrical cantilevered beams with lengths $L = 0.3$ m and radii $r_1 = r_2 = 6$ mm, each loaded with a force $P/2 = 90$ N. The total force is therefore $P = 180$ N (Figure 4.3a).

The combined cross-sectional area of the non-aggregated structure is therefore $\pi r_1^2 + \pi r_2^2 = 226.19 \text{mm}^2$.

The aggregated structure is a single beam with the same length L and a reduced cross-sectional area with the factor $2^{1/3}$ (1.26). It carries the same total load of

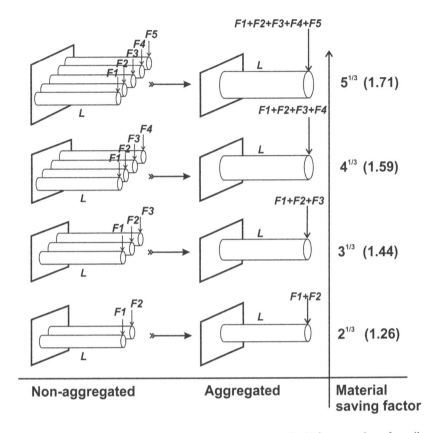

FIGURE 4.2   Materials saving factors achieved by the method of aggregation of cantilever beams.

180 N (Figure 4.3a). Therefore, for the cross-sectional area of the aggregated beam, we have: $\pi R^2 = 226.19/2^{1/3}$, from which

$$R = \sqrt{226.19 / \left(2^{1/3} \times \pi\right)} = 7.559 \text{ mm} \qquad (4.9)$$

where R is the radius of the single aggregated beam.

   In the second variant of aggregation, the non-aggregated structure is composed of three cylindrical cantilevered beams with lengths $L = 0.3$ m and radii $r_1 = r_2 = r_3 = 6$mm, each loaded with a force $180/3 = 60$ N (Figure 4.3b). The total loading force is $P = 180$ N and the combined cross-sectional area of the non-aggregated structure is $\pi r_1^2 + \pi r_2^2 + \pi r_3^2 = 339.292$mm^2. The aggregated structure is a single beam with the same length L and a reduced cross-sectional area with the factor $3^{1/3}$ (1.44). It supports the same total load of 180 N (Figure 4.3b).

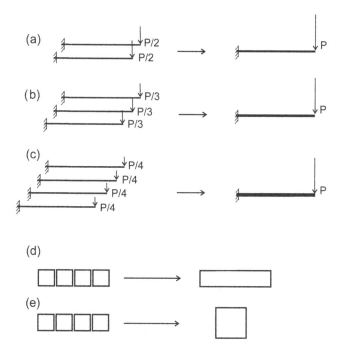

FIGURE 4.3 (a) Two-to-one, b) three-to-one and (c) four-to-one aggregation of cantilever beams; (d) aggregation into a geometrically dissimilar cross-section; (e) aggregation into a geometrically similar cross-section.

Therefore, for the cross-sectional area of the single beam we have: $\pi R^2 = 339.292/3^{1/3}$, from which

$$R = \sqrt{339.292/\left(3^{1/3} \times \pi\right)} = 8.653 \text{ mm}, \qquad (4.10)$$

where R is the radius of the single aggregated beam.

In the third variant of aggregation, the non-aggregated structure is composed of four cylindrical cantilevered beams with lengths $L = 0.3$m and radii $r_1 = r_2 = r_3 = r_4 = 6$mm, each loaded with a force $180/4 = 45$ N (Figure 4.3c). The total loading force is therefore again $P = 180$ N. The combined cross-sectional area of the non-aggregated structure is $\pi r_1^2 + \pi r_2^2 + \pi r_3^2 + \pi r_4^2 = 452.389$ mm^2. The aggregated structure is a single beam which carries the same load of 180 N, has the same length L and a reduced cross-sectional area with the factor $4^{1/3}$ (1.587). Therefore, $\pi R^2 = 452.389/4^{1/3}$, from which

$$R = \sqrt{452.389/\left(4^{1/3} \times \pi\right)} = 9.524 \text{ mm} \qquad (4.11)$$

is obtained, where R is the radius of the single aggregated beam.

TABLE 4.1

Results for the Maximum Tensile Stress for the Non-Aggregated and Aggregated Cantilever Beams

	Non-Aggregated	Aggregated
Two-to-one	159.2 MPa	159.2 MPa
Three-to-one	106.1 MPa	106.1 MPa
Four-to-one	79.6 MPa	79.6 MPa

The maximum tensile stress from loading the cantilevered beams has been calculated by using the classical stress analysis formula (4.4) for simple bending and the results have been listed in Table 4.1. As can be seen from the table, *the maximum tensile stress is the same for the aggregated and non-aggregated structures, which demonstrates that at the same maximum tensile stress, the aggregated structure can support the same total load with a much smaller volume of material* (Todinov, 2024a). The material saving factor s_f is given by the Equation (4.8).

4.3.3 LOAD-BEARING CAPACITY COMPARISON BETWEEN TWO CANTILEVERED STRUCTURES BUILT WITH THE SAME TOTAL VOLUME OF MATERIAL

Consider again two alternative structures, each including m and n identical load-bearing beams, correspondingly, where $m < n$ and the length L of the beams in both structures is the same. The cross-sectional area of each beam from the first (aggregated) structure is S_1 while the cross-sectional area of each beam from the second (non-aggregated) structure is S_2 and $S_1 > S_2$. This means that the aggregated structure has been obtained from the non-aggregated structure by consolidating the beams into fewer beams with larger cross sections (Figure 4.1a).

The larger cross sections of the beams in the aggregated structure are geometrically similar to the smaller cross sections of the beams in the non-aggregated structure. For example, if the non-aggregated section is a circular or square cross section, the aggregated cross section is also a circular or square cross section. In addition, the volume of material for both structures is the same which means that the relationship

$$mS_1L = nS_2L \qquad (4.12)$$

holds, from which

$$S_1 / S_2 = n / m \qquad (4.13)$$

If the radius of the load-bearing beams from the aggregated structure is r_1 and that for the non-aggregated structure is r_2, from equation (4.13), it follows that

$$r_1 / r_2 = \sqrt{n / m} \qquad (4.14)$$

Suppose that the maximum tensile stress permitted by the material of the beams is σ_{cr}. Let P_1 denote the total load the aggregated structure can support to a critical stress of magnitude σ_{cr}. Let P_2 denote the total load that the non-aggregated structure can support to a critical stress of magnitude σ_{cr}. It is assumed that the beams are loaded uniformly, which means that each beam from the aggregated structure supports a load of magnitude P_1/m while each beam from the non-aggregated structure supports a load of magnitude P_2/n.

Let us compare the load-bearing capacities of the aggregated and non-aggregated structure *built with the same total volume of material*. Since for a circular cross section, the second moment of area is $I = \pi r^4/4$, expression (4.4) yields

$$\sigma = 4M / \left(\pi r^3\right) = 4PL / \left(\pi r^3\right) \tag{4.15}$$

where P is the loading force and L is the length of the cantilever beam. If $\sigma = \sigma_{cr}$ is set in (15) and the equation is solved with respect to the loading force P, an expression for the maximum loading force that sets a critical tensile stress of magnitude σ_{cr} is obtained:

$$P = \sigma_{cr}\pi r^3 / \left(4L\right) \tag{4.16}$$

Let P_1 be the maximum total load that the aggregated structure composed of m beams with radii r_1, can support. Since the loading per beam from the aggregated structure is P_1/m, substituting in (4.16) gives

$$P_1 / m = \sigma_{cr}\pi r_1^3 / \left(4L\right) \tag{4.17}$$

Similarly, let P_2 be the maximum total load that the non-aggregated structure composed of n beams with radii r_2 can support. Since the loading per beam from the non-aggregated structure is P_2/n, substituting in (4.16) gives

$$P_2 / n = \sigma_{cr}\pi r_2^3 / \left(4L\right) \tag{4.18}$$

Let $x_1 = x_2 = \ldots = x_m = \left(\pi r_1^2 L\right)$ stand for the volumes of the load-bearing beams from the aggregated structure and $y_1 = y_2 = \ldots = y_n = \left(\pi r_2^2 L\right)$ stand for the volumes of the load-bearing beams from the non-aggregated structure. Equation (4.17) can then be written as

$$P_1 / m = \left[\pi\sigma_{cr} / \left(4L\right)\right] \times \left[x_1^{3/2} / \left(\pi^{3/2} L^{3/2}\right)\right]$$

while Equation (4.18) can be written as:

$$P_2 / n = \left[\pi \sigma_{cr} / (4L) \right] \times \left[y_2^{3/2} / \left(\pi^{3/2} L^{3/2} \right) \right]$$

Next, denote $a_1 = \ldots = a_m = \ldots = a_n = a = \pi \sigma_{cr}/(4L\pi^{3/2}L^{3/2})$. The sufficient conditions (4.2) for the super-additive inequality (4.3) are then fulfilled and according to inequality (4.3):

$$ax_1^{3/2} + \ldots + ax_m^{3/2} > ay_1^{3/2} + \ldots + ay_n^{3/2} \qquad (4.19)$$

The left-hand side of inequality (4.19), $P_1 = ax_1^{3/2} + \ldots + ax_m^{3/2}$, is the total bending load supported by the aggregated structure while the right-hand side $P_2 = ay_1^{3/2} + \ldots + ay_n^{3/2}$ is the total bending load supported by the non-aggregated structure. Considering that the left-hand side of (4.19) is equal to $P_1 = ma\left(\pi r_1^2 L \right)^{3/2}$ and the right-hand side is equal to $P_2 = na\left(\pi r_2^2 L \right)^{3/2}$, the ratio of the two forces can be obtained:

$$P_1 / P_2 = \left(m / n \right) \times \left(r_1 / r_2 \right)^3$$

Since the volume of material used for building both structures is the same, from equation (4.14) it follows that

$$\left(r_1 / r_2 \right)^3 = \left(\sqrt{n / m} \right)^3 = \left(n / m \right)^{3/2}$$

and the ratio P_1/P_2 of the load-bearing capacities of the aggregated and non-aggregated structure becomes

$$P_1 / P_2 = \left(m / n \right) \times \left(n / m \right)^{3/2} = \sqrt{n / m} \qquad (4.20)$$

For $m = 2$ and $n = 8$, for example, this ratio is $P_1/P_2 = 2$ which demonstrates that for the same total volume of the load-bearing beams in the two structures, the aggregated structure (Figure 4.1a) has a significantly larger load-bearing capacity than the non-aggregated structure.

It is essential that the larger cross-sections of the beams in the aggregated structure are geometrically similar to the smaller cross-sections of the beams from the non-aggregated structure. If this requirement is not met, the load-carrying capacity of the aggregated structure may not increase.

Thus, for the same cross-sectional area of non-aggregated and aggregated structures, the larger, geometrically dissimilar cross-section (shown in Figure 4.3d) does not improve the load-bearing capacity. Conversely, the geometrically similar cross-section (as shown in Figure 4.3e) does enhance the load-bearing capacity of the non-aggregated structure.

4.3.4 LOAD-BEARING CAPACITIES AT THE SAME MAXIMUM PERMISSIBLE DEFLECTION OF TWO CANTILEVERED STRUCTURES BUILT WITH THE SAME TOTAL VOLUME OF MATERIAL

Suppose now that the limiting condition during loading is not the maximum permissible tensile stress but the maximum permissible deflection δ_{cr}. Let the loads that the aggregated and non-aggregated structures can support at a critical deflection of magnitude δ_{cr} be P_1 and P_2, correspondingly. The load-bearing beams are loaded uniformly. This means that each beam from the aggregated structure supports a load of magnitude P_1/m while each beam from the non-aggregated structure supports a load of magnitude P_2/n.

Because the volume of material for both structures is the same, the ratio of the radii is given by Equation (4.14). For a cantilever beam, the link between a deflection of magnitude δ_{cr} and the loading force P that causes this deflection is given by the classical formula (Gere and Timoshenko, 1999; Beer et al., 2002):

$$\delta_{cr} = PL^3 / (3EI) \tag{4.21}$$

where E is the Young's modulus of the material and I is the second moment of area. Solving (4.21) for P gives:

$$P = 3EI\delta_{cr} / L^3 \tag{4.22}$$

Considering that for a circular section, $I = \pi r^4/4$, equation (4.22) becomes:

$$P = 3E\pi r^4 \delta_{cr} / \left(4L^3\right) \tag{4.23}$$

Since each beam from the aggregated structure has a radius r_1 and is loaded by a force P_1/m, for a single beam from the aggregated structure, the following relationship holds:

$$P_1 / m = 3E\pi r_1^4 \delta_{cr} / \left(4L^3\right) \tag{4.24}$$

For the non-aggregated structure, each beam has a radius r_2 and is loaded by a force P_2/n. Therefore, for a single beam from the non-aggregated structure, the corresponding relationship is:

$$P_2 / n = 3E\pi r_2^4 \delta_{cr} / \left(4L^3\right) \tag{4.25}$$

It can be shown that the inequality: $P_1 > P_2$ holds if $n > m$. Indeed, taking the ratio of (4.24) and (4.25) gives

$$P_1 n / \left(P_2 m\right) = r_1^4 / r_2^4 \tag{4.26}$$

From relationship (4.14), $r_1^4 / r_2^4 = n^2 / m^2$ is obtained and the substitution in (4.26) gives

$$P_1 n / (P_2 m) = r_1^4 / r_2^4 = n^2 / m^2$$

from which

$$P_1 / P_2 = n / m \qquad\qquad (4.27)$$

Again, the aggregation of the cross sections led to an increased load-bearing capacity of the aggregated structure despite that both structures are built with the same volume of material. For $n = 8$, $m = 2$, for example, the aggregated structure has four times greater load-bearing capacity compared to the non-aggregated structure:

$$P_1 / P_2 = n / m = 4 \qquad\qquad (4.28)$$

Although the load per beam increases in the aggregated structure, this is outweighed by the increase in the second moment of area of the cross section. For cantilevered beams, the deformation is given by Equation (4.21). Despite the increase of the load P in the numerator from aggregating the loads, this increase is outweighed by the more significant increase of the second moment of area 'I' of the cross sections, caused by the increased radius r. Indeed, the second moment of area depends on the fourth power of the radius of the beam ($I = \pi r^4 / 4$).

The aggregation method, can indeed be derived by bypassing the sub-additive and super-additive inequalities (4.1) and (4.3). However, this can only be accomplished for uniform loading and uniform cross sections. In cases of a non-uniform loading and non-uniform cross sections, the general form of inequalities (4.1) and (4.3) must be used. It is not clear how the results related to non-uniform loading or non-uniform cross sections can be derived by bypassing the inequalities (4.1) and (4.3).

4.4   STRUCTURES BUILT ON SIMPLY SUPPORTED BEAMS LOADED IN BENDING

Consider now two alternative structures including m and n load-bearing simply supported beams, correspondingly, where $m < n$ (Figure 4.4a and b). Again, the length L of the beams in both structures is the same. The structure in Figure 4.4a, consisting of fewer (m) load-bearing beams with larger cross-sectional area will be referred to as 'aggregated' structure while the structure in Figure 4.4b, consisting of a larger number (n) of load-bearing beams with smaller cross-sectional area, will be referred to as 'non-aggregated' structure.

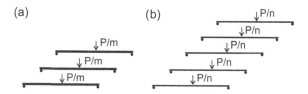

FIGURE 4.4 (a) Aggregated and (b) non-aggregated structure built on simply supported beams.

4.4.1 MINIMUM VOLUME OF MATERIAL NEEDED TO CARRY A LOAD OF A SPECIFIED MAGNITUDE

Similar to the case of the cantilevered beams, suppose that σ_{cr} is the maximum permissible stress of the material for the structures in Figure 4.4. Assume a uniform distribution of the load over the load-bearing beams. The loads per beam for the aggregated and non-aggregated structures are P/m and P/n, respectively.

Using a very similar reasoning to the reasoning in Section 4.3.2 related to cantilevered beams (which will not be repeated here), the ratio of the volumes of aggregated and non-aggregated structures necessary to support a total load of magnitude P can be determined. The ratio of the minimum volumes necessary to support a total load of magnitude P is again given by expression (4.7) ($V_1/V_2 = (n/m)^{1/3}$) and for $m = 1$, the material saving factor $s_f = V_1/V_2 = n^{1/3}$ is given by expression (4.8). Again, the total volume V_1 of the load-bearing beams in the aggregated structure needed to support the total load P is significantly smaller than the total volume V_2 of the load-bearing beams in the non-aggregated structure needed to support the same total load P.

4.4.2 COMPARISON AND VERIFICATION

The comparison of structures involving simply supported beams also involves three types of aggregation. First, two beams carrying a total load P were considered, aggregated into a single beam with the same length L and with a $2^{1/3}$ times smaller volume, carrying the same total load P (Figure 4.5a). Next, three beams carrying a total load P were considered, aggregated into a single beam with the same length L and a $3^{1/3}$ times smaller volume, supporting the same total load P (Figure 4.5b). Finally, four beams carrying a specified total load P were considered, aggregated into a single beam with the same length L and a $4^{1/3}$ times smaller volume, supporting the total load P (Figure 4.5c).

In the first variant of aggregation (Figure 4.5a), the non-aggregated structure is composed of two simply supported cylindrical rods with lengths $L = 0.8$ m and radii $r_1 = r_2 = 6$ mm, each loaded with a force $P/2 = 180N$. The total force is therefore $P = 360N$. The aggregated beam is with the same length L, has a reduced cross-sectional area with the factor $2^{1/3}$ (1.26) and it supports the same total load

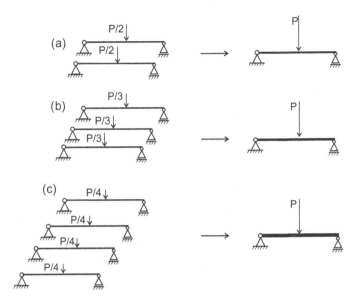

FIGURE 4.5 (a) Two-to-one, (b) three-to-one and (c) four-to-one aggregation of simply supported beams.

of 360 N. The radius $R = 7.559$ mm of the single aggregated beam was calculated with Equation (4.9) also used for calculating the cross section of the aggregated cantilever beam.

In the second variant of aggregation (Figure 4.5b), the non-aggregated structure was composed of three simply supported beams with lengths $L = 0.8$ m and radii $r_1 = r_2 = r_3 = 6$ mm, each loaded with a force $P/3 = 360/3 = 120$ N. The total loading force is therefore again $P = 360$ N. The radius $R = 8.653$ mm of the single aggregated beam was calculated using Equation (4.10), also used for calculating the cross section of the aggregated cantilever beam.

Finally, in the third variant of aggregation (Figure 4.5c), the non-aggregated structure was composed of four cylindrical simply supported beams with lengths $L = 0.8$ m and radii $r_1 = r_2 = r_3 = r_4 = 6$ mm, each loaded with a force $P/4 = 360/4 = 90$ N. The total loading force is therefore $P = 360$ N. The radius $R = 9.524$ mm of the single aggregated beam was calculated using Equation (4.11), also used for calculating the cross section of the aggregated cantilever beam. The maximum tensile stress from loading the simply supported structures has been calculated by using the standard stress analysis formula (4.4) for simple bending and the results are listed in Table 4.2.

As can be seen from Table 4.2, the maximum tensile stress is the same for the aggregated and non-aggregated structures. This demonstrates that at the same maximum tensile stress, the aggregated structure can support the same total load P with a much smaller volume of material (Todinov, 2024a).

TABLE 4.2

Maximum Tensile Stress for the Non-Aggregated and Aggregated Simply Supported Beams

	Non-Aggregated	Aggregated
Two-to-one	212.2 MPa	212.2 MPa
Three-to-one	141.5 MPa	141.5 MPa
Four-to-one	106.1 MPa	106.1 MPa

4.4.3 Load-Bearing Capacity of Simply Supported Beam Structures Built with the Same Total Volume of Material

Suppose that the permissible stress of the material for the beams is σ_{cr}. Let P_1 be the load that the aggregated structure can support, which corresponds to the permissible stress σ_{cr}. Let P_2 be the load that the non-aggregated structure can support, which corresponds to the permissible stress σ_{cr}. The load-bearing beams are loaded uniformly. This means that each beam from the aggregated structure in Figure 4.4a carries load of magnitude P_1/m while each beam from the non-aggregated structure in Figure 4.4b carries load of magnitude P_2/n. Using a very similar reasoning to the reasoning in Section 4.3.2 related to cantilevered beams (which will not be repeated here), the ratio of the load-bearing capacities of the aggregated and non-aggregated structure at the same total volume of material used for building the structures can be determined. The ratio is given with the equation:

$$P_1 / P_2 = \sqrt{n / m} \qquad (4.29)$$

which is the same equation as the one derived for cantilevered beams. For $m = 2$ and $n = 8$, for example, Equation (4.29) gives: $P_1/P_2 = 2$. Again, for the same total volume of the load-bearing beams in the two structures, the aggregated structure (Figure 4.4a) has a load-bearing capacity two times larger than that of the non-aggregated structure (Figure 4.4b).

4.4.4 Load-Bearing Capacity at the Same Maximum Permissible Deflection for Structures Built with the Same Total Volume of Material

Suppose that the limiting condition for the loading is not the maximum tensile stress but the maximum permissible deflection. Let the maximum permissible deflection δ_{cr} be the same for both the non-aggregated and aggregated structures. Suppose that P_1 is the total load on the aggregated structure in Figure 4.4a that corresponds to a maximum permissible deflection of magnitude δ_{cr}. Similarly, P_2 is the total load on the non-aggregated structure in Figure 4.4b that corresponds to a maximum permissible deflection of magnitude δ_{cr}. The load-bearing beams are loaded uniformly. This means that each beam from the aggregated structure in Figure 4.4a carries load of magnitude P_1/m while each beam from the non-aggregated structure in Figure 4.4b carries load of magnitude P_2/n. Because the volume of material for both structures is the same, the ratio of the radii of the beams is given by equation (4.14). The link between the maximum permissible deflection δ_{cr} and the loading force P that causes this deflection for a simply-supported beam is given by (Gere and Timoshenko, 1999; Beer et al., 2002):

$$\delta_{cr} = PL^3 / (48EI)$$

from which

$$P = 48EI\delta_{cr} / L^3 \tag{4.30}$$

Considering that for a circular section, $I = \pi r^4/4$, equation (4.30) becomes:

$$P = 12E\pi r^4 \delta_{cr} / L^3 \tag{4.31}$$

Each beam from the aggregated structure in Figure 4.4a has a radius r_1 and is loaded by a force P_1/m. Therefore, for a single beam from the aggregated structure, the following relationship holds:

$$P_1 / m = 12E\pi r_1^4 \delta_{cr} / L^3 \tag{4.32}$$

Similarly, each beam from the non-aggregated structure in Figure 4.4b has a radius r_2 and is loaded by a force P_2/n. Therefore, for a single beam from the non-aggregated structure, the relationship is:

$$P_2 / n = 12E\pi r_2^4 \delta_{cr} / L^3 \tag{4.33}$$

Taking the ratio of (4.32) and (4.33) gives

$$P_1 n / (P_2 m) = r_1^4 / r_2^4 \tag{4.34}$$

From relationship (4.14), $r_1^4 / r_2^4 = n^2 / m^2$ and the substitution in (4.34) gives

$$P_1 n / P_2 m = r_1^4 / r_2^4 = n^2 / m^2$$

and the ratio of the load-bearing capacities becomes:

$$P_1 / P_2 = n / m \tag{4.35}$$

This result is identical with the ratio of the load-bearing capacities (4.27) obtained for structures built on cantilevered beams. The aggregation of the cross sections of simply-supported beams leads to an increased load-bearing capacity of the structure at the same volume of material.

4.5 LOAD-BEARING CAPACITY COMPARISON BETWEEN Γ-FRAMES BUILT WITH THE SAME TOTAL VOLUME OF MATERIAL

The Γ-frames are examples of more complex structures loaded in bending. The type of aggregation selected for the Γ-frames was also 8 to 2. A non-aggregated structure containing 8 load-bearing elements (Figure 4.6a) with diameters $\phi 1$ mm has been aggregated into a structure containing only 2 load-bearing elements with diameters $\phi 2$ mm (Figure 4.6b). The other dimensions of the Γ-frames are according to Figure 4.6. The diameters of the Γ-frames from the aggregated and non-aggregated structure were selected such that the total volume of material used for the aggregated and non-aggregated structures to be the same.

The loading force applied to each of the non-aggregated Γ-frames in Figure 4.6a is $P/8$ while the loading force applied to each of the aggregated Γ-frames in Figure 4.6b is $P/2$. The total loading force for both structures was the same – equal to $P = 24$ N and the Young modulus of the material for the frames was chosen to be 210 GPa. Table 4.3 summarises the results related to the maximum tensile stress and deflections related to the non-aggregated and aggregated Γ-frames. All calculations have been produced through standard techniques from stress analysis and to conserve space, details have been omitted.

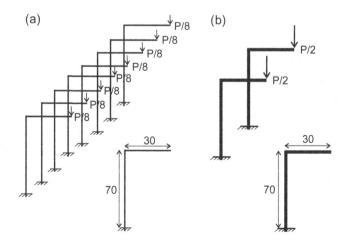

FIGURE 4.6 Eight-to-two aggregation of loaded Γ-frames;(a) non-aggregated structure; (b) aggregated structure.

TABLE 4.3

Results for the Maximum Stress and Deflection for the Non-Aggregated and Aggregated Γ-Frames

	Non-Aggregated	Aggregated
Maximum stress	920.56 MPa	462.18 MPa
Maximum deflection	20.95 mm	5.23 mm

From the results for the loaded Γ-frames in Table 4.3, we can conclude that for the same volume of material, the aggregated structure is characterised with significantly smaller maximum stress and deflection. This is another confirmation of the power of the aggregation method.

4.6   EXPERIMENTAL VERIFICATION

In all conducted experiments (Todinov, 2024a), a digital dynamometer capturing the applied load was used, with a range (0–50 N) and measurement accuracy of 1%. The displacement was captured by using a digital depth gauge with a measurement range 0–80 mm and measurement precision 0.01 mm. In all conducted experiments, the experimental verification of the aggregation method was conducted by using wire with diameters of 1 mm and 2 mm and Young's modulus 179 GPa. The wire diameters were selected in such a way that the total cross-sectional area of two wires of diameter 2 mm is exactly equal to the total cross-sectional area of eight wires of diameter 1 mm:

$$2 \times \frac{\pi \times 2^2}{4} = 8 \times \frac{\pi \times 1^2}{4} = 2\pi \ \mathrm{mm}^2$$

4.6.1   Configuration of the Experimental Equipment for Cantilevered Structures

The experimental verification of the proposed aggregation method on cantilevered beams was conducted by using rods of wire with diameters of 1 mm and 2 mm. A cantilever beam, labelled as '1' (Figure 4.7) and made of wire was securely anchored in the vice, labelled as '2'. The beam's deflection was controlled by a digital depth gauge, labelled as '4', while the load applied to the free end of the beam was quantified using a digital dynamometer, labelled as '3'. The length L of the cantilever beam was 40 mm for both wire diameters. The beam deflection was specified to be exactly 2 mm and the loading force at this deflection was recorded. During the experiments, all deformations remained in the elastic region. The position of the digital gauge '4' can be easily adjusted by moving it sideways along the ferromagnetic rail '5'. The permanent magnets in the legs of the digital gauge ensure strong contact with the rail and a stable position during measurements.

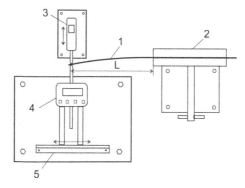

FIGURE 4.7   Configuration of the experimental equipment for cantilever beams.

4.6.2   Experimental Results for Cantilever Beams

For the wire of 2 mm diameter, the average recorded force to reach 2mm deflection was 13.16 N while for the wire of 1 mm diameter, the average recorded force to reach 2 mm deflection was 0.85 N. Consequently, if uniformly loaded, two beams of 2 mm diameter will carry a total load of $2 \times 13.16 = 26.32N$ while 8 beams of 1 mm diameter will carry a total load of $8 \times 0.85 = 6.8N$. As a result, at the same total volume and at the same specified deflection, the aggregated structure carries a significantly larger load than the non-aggregated one.

4.6.3   Configuration of the Experimental Equipment for Simply Supported Beams

The experimental verification of the proposed aggregation method on simply supported beams was conducted using rods of wire with diameters of 1mm and 2 mm. The simply supported beam labelled as '1' was securely positioned on the

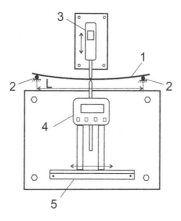

FIGURE 4.8   Configuration of the experimental equipment for simply-supported beams.

TABLE 4.4

Maximum Load [N] Until 2 mm Deflection for 8 Non-Aggregated Versus 2 Aggregated Cantilever and Simply-Supported Beams

	8 Non-Aggregated Beams	2 Aggregated Beams
Cantilever beams	6.8 N	26.32 N
Simply-supported beams	13.44 N	52.8 N

rolling supports '2' (Figure 4.8). The deflection of the simply-supported beam' 1' was measured by a digital depth gauge '4', while the load applied in the middle of the simply-supported beam '1' was measured by the digital dynamometer '3'. The distance between the supports '2' was 80mm for both diameters of the wire and the concentrated load from the push-dynamometer 3 was applied in the middle of the beam, at a distance of 40mm from the supports '2'. Again, the beam deflection in the middle was specified to be exactly 2mm and the loading force at this deflection was recorded. During the experiment, all deformations remained in the elastic region.

4.6.4 EXPERIMENTAL RESULTS FOR SIMPLY-SUPPORTED BEAMS

For the wire of 1 mm diameter, the average recorded force to reach 2 mm deflection in the middle was 1.68 N while for the wire of 2 mm diameter, the average recorded force was 26.4 N.

Consequently, two uniformly loaded beams of 2 mm diameter will support a total load of $2 \times 26.4 = 52.8$ N while eight uniformly loaded beams of 1 mm diameter will support a total load of $8 \times 1.68 = 13.44$ N. The results from the experimental study related to cantilevered and simply supported beams have been summarised in Table 4.4 and confirm that aggregation significantly increases the load-bearing capacity of structures built on simply supported beams subjected to bending.

4.6.5 CONFIGURATION OF THE EXPERIMENTAL EQUIPMENT FOR Γ-FRAMES

The experimental verification of the proposed aggregation method on Γ-frames was also conducted using frames made of wire with diameters of 1 mm and 2 mm. Again, the selected aggregation was 8-to-2 and the dimensions of the frames (in mm) are according to Figure 4.9.

A Γ-frame, labelled as '1' was securely anchored in the vice '2' (Figure 4.9). The frame's deflection was controlled by a digital depth gauge '4', while the load applied was measured using the digital dynamometer '3'.

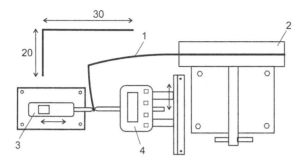

FIGURE 4.9 Experimental verification of the aggregation method by using Γ-frames.

TABLE 4.5
Maximum Deflection for 8 Non-Aggregated
Γ-Frames Versus 2 Aggregated Γ-Frames

	8 Non-Aggregated Γ-frames	2 Aggregated Γ-frames
Deflection	1.59 mm	0.4 mm

4.6.6 EXPERIMENTAL RESULTS FOR Γ-FRAMES

The selected total bending load on the aggregated and non-aggregated Γ-frame was the same – equal to 8 N and the deflection at this total load was measured. Because of the uniform loading, each element of the non-aggregated structure (8 frames) is effectively loaded with 1 N force while each element of the aggregated structure (2 frames) is loaded with 4 N ($8 \times 1\,N = 2 \times 4\,N$). The measured deflections from the experiments with the Γ-frames have been summarised in Table 4.5. Based on the conducted measurements, it can be concluded that the deflections of the aggregated structure are significantly smaller than those of the non-aggregated structure. This confirms that aggregation increases the load-bearing capacity of structures subjected to bending.

4.6.7 APPLICATIONS

Here is a list of possible applications of load-bearing elements loaded in bending that can benefit from aggregation:

- Steel beams supporting a distributed load, such as a floor in a building.
- Wooden joists arranged side by side to support the weight of a flooring system.
- Console I-beams installed in parallel to bear the load of a structure.
- I-beams installed in parallel to bear the load of a walkway or bridge.

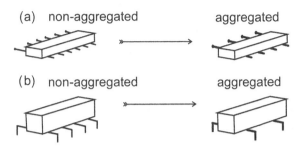

FIGURE 4.10 An application example of non-aggregated and aggregated structures: (a) a supporting structure built on console arms; (b) a supporting structure built on Π-frames.

- Reinforced concrete beams supporting the weight of a suspended parking deck.
- Parallel frames carrying the load of a prefabricated modular structure.
- Deck joists arranged uniformly to provide support for an outdoor decking system.

Simple application examples of non-aggregated and aggregated supporting structures are shown in Figure 4.10a and b, respectively. The six-to-four aggregation in Figure 4.10a and the four-to-two aggregation in Figure 4.10b consists of reducing the number of loaded elements and increasing their cross-sectional area accordingly. This operation substantially reduces the volume of material required to construct the supporting structures.

It needs to be pointed out that in conducting aggregation, the structural safety considerations should always apply. If failure of any of the aggregated elements induces collapse of the entire structure and the consequences of structural collapse are severe, the non-aggregated structure may be safer than the aggregated one.

The results reported in this paper are related to statically determinate load-carrying elements with circular cross-sections loaded in bending. Simulations were also carried out (to be published elsewhere) with statically indeterminate elements and elements with square cross sections loaded in bending. The obtained results confirmed the power of the proposed method of aggregation.

5 Reliability-Related Reverse Engineering of Algebraic Inequalities

An application of the method of algebraic inequalities to improve reliability and reduce risk has been demonstrated in (Todinov, 2020b) where algebraic inequalities have been used to rank the reliabilities of systems with unknown reliabilities of their components. This approach can be summarised as follows. For each of the two competing alternatives 1 and 2 of a system, a reliability network is built first. Next, by using methods from system reliability analysis, the system reliabilities R_1 and R_2 of the competing alternatives are determined. The final step is trying to prove one of the inequalities $R_1 - R_2 > 0$ or $R_2 - R_1 > 0$ (irrespective of the specific reliabilities of the components), which demonstrates the superior system reliability of one of the alternatives. We need to point out here that the direct approach of the method of algebraic inequalities has a limitation: is is not guaranteed to work for all compared alternatives. This means that in some cases it is not possible to prove any of the of the inequalities $R_1 - R_2 > 0$ or $R_2 - R_1 > 0$, irrespective of the reliabilities of the components building the systems.

The question on optimal redundancy allocation for systems with series arrangement of independently working components has been considered in (Valdes and Zequeira, 2006). The results obtained however, have limited validity because they are related to very simple systems, incorporating only two-components.

No publications exist that are related to generating new knowledge by interpreting non-trivial algebraic inequalities, which is subsequently used for improving the reliability of systems or processes.

5.1 RELIABILITY OF SYSTEMS WITH COMPONENTS LOGICALLY ARRANGED IN SERIES AND PARALLEL

Before introducing reverse engineering of algebraic inequalities with the purpose of improving system reliability, the basics of evaluating the reliability of systems, with components logically arranged in series and parallel, will be covered first.

Consider a system including n independently working components. Let S denote the event 'the system is in working state at the end of a specified time interval' and C_k ($k = 1, 2, ..., n$) denote the events 'component k is in working state at the end of the specified time interval'. For components logically arranged in series (Figure 5.1a), the system is in working state at the end of the specified time interval only if all components are in working state at the end of the time interval.

(a)   (b)

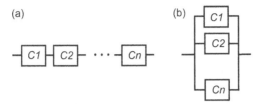

FIGURE 5.1   (a) A system with components (a) logically arranged in series; (b) logically arranged in parallel.

Reliability is the ability of an entity to work without failure for a specified time interval, under specified conditions and environment. The ability to work without failure within the specified time interval is measured by the probability of working without failure during that interval.

According to the reliability theory (Bazovsky, 1961; Hoyland and Rausand, 1994; Ebeling, 1997), the probability of system success (system in working state at the end of the specified time interval) is a product of the probabilities that the components will be in working state at the end of the specified time interval:

$$P(S) = P(C_1) \times P(C_2) \times \ldots \times P(C_n) \qquad (5.1)$$

Denoting by R the probability $P(S)$ that the system will be in working state at the end of the specified time interval and by $R_k = P(C_k)$ the probability that the kth component will be in working state at the end of the specified time interval, Equation (5.1) becomes

$$R = R_1 \times R_2 \times \ldots \times R_n \qquad (5.2)$$

In Equation (5.2), R will be referred to as the reliability of the system and R_k as the reliability of the kth component.

Now consider independently working components logically arranged in parallel (Figure 5.1b). According to the system reliability theory (Bazovsky, 1961; Hoyland and Rausand, 1994; Ebeling, 1997), the probability of system success (system in working state) is equal to the probability that at least a single component will be in working state at the end of the specified time interval.

The event 'at least a single component will be in working state at the end of the specified time interval' and the event 'none of the components will be in working state at the end of that interval' are complementary events whose probabilities add up to unity. Therefore, the probability that at least a single component will be in working state can be evaluated by subtracting from unity the probability that none of the components will be in working state. The advantage offered by this inverse-thinking approach is that the probability that none of the components will be in working state at the end of the specified time interval is easy to calculate.

Indeed, if $R_1, R_2, \ldots, R_n$ denote the reliabilities of the separate components, the probability $P(\bar{S})$ that none of the components will be in working state at the end of the specified time interval (the probability of system failure) is given by

$$P(\bar{S}) = (1 - R_1)(1 - R_2)\ldots(1 - R_n) \tag{5.3}$$

Consequently, the probability $P(S)$ that the system will be in working state is given by

$$P(S) = 1 - P(\bar{S}) = R = 1 - (1 - R_1)(1 - R_2)\ldots(1 - R_n) \tag{5.4}$$

Note that for a logical arrangement of the components in series, the system reliability is a product of the reliabilities of the components, while for a logical arrangement of the components in parallel, the probability of system failure is a product of the probabilities of failure of the components.

A system with components logically arranged in series and parallel can be simplified in stages, as shown in Figure 5.2. In the first stage, the components in Figure 5.2a, logically arranged in series, with reliabilities R_1, R_2 and R_3 are reduced to an equivalent component with reliability $R_{123} = R_1 R_2 R_3$. The components logically arranged in series with reliabilities R_4 and R_5 are reduced to an equivalent component with reliability $R_{45} = R_4 R_5$ and the components from the parallel branch, with reliabilities R_6 and R_7 are reduced to an equivalent component with reliability $R_{67} = R_6 R_7$. The resultant equivalent network is shown in Figure 5.2b.

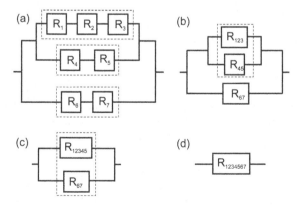

FIGURE 5.2 Network reduction method for determining the reliability of a system including components logically arranged in series and parallel (a) initial network; (b) the network after the first network reduction; (c) the network after the second network reduction; (d) the original network reduced to a single component.

In the second stage, the components in parallel, with reliabilities R_{123} and R_{45} in Figure 5.2b are reduced to an equivalent component with reliability $R_{12345} = 1 - (1 - R_{123})(1 - R_{45})$. The resultant equivalent reliability network is shown in Figure 5.2c.

Finally, the reliability network in Figure 5.2c is further simplified by reducing equivalent components with reliabilities R_{12345} and R_{67} to a single equivalent component with reliability $R_{1234567} = 1 - (1 - R_{12345}) \times (1 - R_{67})$ (Figure 5.2d). This is also the final result for the reliability of the original network in Figure 5.2a.

This is the essence of the *network reduction method* for determining the reliability of series-parallel systems where components are logically arranged only in series and parallel (Ebeling, 1997). Various techniques for determining system reliability of complex networks that are not with series-parallel topology have been considered in (Todinov, 2016).

5.2 IMPROVING THE RELIABILITY OF A SYSTEM WITH INTERCHANGEABLE REDUNDANCIES BY REVERSE ENGINEERING OF ALGEBRAIC INEQUALITIES

5.2.1 REDUNDANCY OPTIMISATION BASED ON KNOWN VALUES OF THE COMPONENT RELIABILITIES

Despite the extensive research on redundancy optimization in the reliability literature, no existing technique has yet addressed redundancy optimisation without requiring specific knowledge of component reliabilities. Traditionally, optimal redundancy allocation methods begin with known reliability values for all components and redundancies. For example, Yi Ding et al. (2021) introduced a multi-performance redundancy optimization technique for multi-state systems using a genetic algorithm, relying on known component reliabilities. Similarly, Aqel and Mohamed (2023) addressed the redundancy allocation problem with nature-inspired AI algorithms, such as adaptive particle swarm optimization, which are also based on known component reliabilities.

To address redundancy allocation in large systems, Florin et al. (2020) presented several approaches, including the Lagrange multipliers technique, a pair-wise hill-climbing algorithm, and an evolutionary algorithm. Additionally, Shubin et al. (2020) reviewed system reliability optimization techniques driven by importance measures, which also depended on known component reliabilities.

The key contribution of this chapter is in demonstrating that by leveraging asymmetry, system reliability can be improved without any knowledge of component reliability values (or probabilities of failure). Specifically, it is established that for series-parallel systems, an asymmetric arrangement of interchangeable redundancies consistently results in higher system reliability compared to a symmetric arrangement, regardless of the individual reliability values (or probabilities of failure) of the components.

For decades, the results reported in this chapter eluded system reliability experts, demonstrating that they could not be achieved without employing reverse engineering of algebraic inequalities.

5.2.2 REDUNDANCY OPTIMISATION BASED ON UNKNOWN VALUES OF THE COMPONENT RELIABILITIES

If the variables $a_1, a_2, \ldots, a_n$ entering a correct algebraic inequality are subjected to the constraints $0 \le a_1, a_2, \ldots, a_n \le 1, i = 1, \ldots, n$, they can, for example, be physically interpreted as probabilities of failure of components working independently from one another (Todinov, 2023a). In addition, if the left- or right-hand side of an algebraic inequality is composed of products of terms of the type $(1 - a_i a_j)$, the term $(1 - a_i a_j)$ can be interpreted as the reliability of a section including two components logically arranged in parallel (Figure 5.3a). The product $(1 - a_i a_j)$ $(1 - a_{i+1} a_{j+1}) \ldots (1 - a_r a_s)$ of several such terms can be interpreted as the reliability of a series-parallel system including several sections logically arranged in series, within each of which, the components are logically arranged in parallel (Figure 5.3b).

Accordingly, the left and right-hand side of inequalities including these terms can be interpreted as reliabilities of alternative system configurations. Next, through algebraic inequalities, the intrinsic reliabilities of the alternative configurations can be compared.

To illustrate this approach, consider the simple algebraic inequality:

$$\left(1 - a_1^2\right)\left(1 - a_2^2\right) \le \left(1 - a_1 a_2\right)\left(1 - a_2 a_1\right) \tag{5.5}$$

where $0 \le a_1 \le 1; 0 \le a_2 \le 1$. Proving this inequality is equivalent to proving the equivalent inequality

$$1 - a_1^2 - a_2^2 + a_1^2 a_2^2 \le 1 - 2a_1 a_2 + a_1^2 a_2^2 \tag{5.6}$$

which, in turn can be proved by proving the equivalent inequality

$$a_1^2 + a_2^2 - 2a_1 a_2 \ge 0 \tag{5.7}$$

Inequality (5.7) however, is true because $a_1^2 + a_2^2 - 2a_1 a_2 = \left(a_1 - a_2\right)^2$ is non-negative.

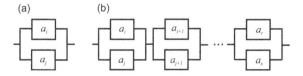

FIGURE 5.3 (a) Sections whose reliabilities are given by products of terms $(1 - a_i a_j)$; (b) Series-parallel systems including sections of type (a).

Inequality (5.5) can be generalised for more than two types of components. Thus, for $n \geq 2$ types of components $A_1, A_2, \ldots, A_n$ with probabilities of failure a_1, $a_2, \ldots, a_n$, correspondingly, inequality (5.5) is generalised to (Todinov, 2023a):

$$\left(1 - a_1^2\right)\left(1 - a_2^2\right)\ldots\left(1 - a_n^2\right) \leq \left(1 - a_1 a_2\right)\left(1 - a_2 a_3\right)\ldots\left(1 - a_n a_1\right) \tag{5.8}$$

Proof: Inequality (5.8) can be proved by induction. For $n = 2$, inequality (5.8) coincides with inequality (5.5) which has been shown to be true.

Now let us assume that inequality (5.8) is true for $n = k$ (induction hypothesis):

$$\left(1 - a_1^2\right)\left(1 - a_2^2\right)\ldots\left(1 - a_k^2\right) \leq \left(1 - a_1 a_2\right)\left(1 - a_2 a_3\right)\ldots\left(1 - a_k a_1\right) \tag{5.9}$$

It can be shown that inequality (5.9) is also true for k+1 number of terms, where k+1th term has been denoted by a_{k+1}.

Without loss of generality, the values $a_1, a_2, \ldots, a_k, a_{k+1}$ can always be ordered in ascending order. In other words, a permutation $a_{1p}, a_{2p}, \ldots, a_{kp}, a_{k+1p}$ of the values $a_1, a_2, \ldots, a_k, a_{k+1}$ can always be found for which $a_{1p} \leq a_{2p} \leq, \ldots, \leq a_{kp} \leq a_{k+1p}$ holds. The reliability of the original system will not be altered because of the permutation because

$$\left(1 - a_1^2\right)\left(1 - a_2^2\right)\ldots\left(1 - a_k^2\right)\left(1 - a_{k+1}^2\right) = \left(1 - a_{1p}^2\right)\left(1 - a_{2p}^2\right)\ldots\left(1 - a_{kp}^2\right)\left(1 - a_{k+1p}^2\right)$$

$$0 \leq a_i \leq 1$$

holds.

According to assumption (5.9), the inequality is always true for $n = k$ terms, therefore we have:

$$\left(1 - a_{1p}^2\right)\left(1 - a_{2p}^2\right)\ldots\left(1 - a_{kp}^2\right) \leq \left(1 - a_{1p} a_{2p}\right)\left(1 - a_{2p} a_{3p}\right)\ldots\left(1 - a_{kp} a_{1p}\right) \tag{5.10}$$

$$0 \leq a_{ip} \leq 1$$

If the correct by assumption inequality (5.10) is multiplied by the non-negative value $\left(1 - a_{k+1p}^2\right)$, the direction of inequality (5.10) will not be altered and the inequality

$$\begin{aligned}(1 - a_{1p}^2)(1 - a_{2p}^2)\ldots(1 - a_{kp}^2)(1 - a_{k+1p}^2) \\ \leq (1 - a_{1p} a_{2p})(1 - a_{2p} a_{3p})\ldots(1 - a_{kp} a_{1p})(1 - a_{k+1p}^2)\end{aligned} \tag{5.11}$$

$$0 \leq a_{ip} \leq 1$$

is obtained. We will show that for the product $\left(1-a_{kp}a_{1p}\right)\left(1-a_{k+1p}^2\right)$ in the right-hand side of (5.11), the following inequality holds:

$$\left(1-a_{kp}a_{1p}\right)\left(1-a_{k+1p}^2\right) \leq \left(1-a_{kp}a_{k+1p}\right)\left(1-a_{k+1p}a_{1p}\right) \tag{5.12}$$

Indeed, expanding the left- and right-hand side of (5.12) gives

$$1-a_{k+1p}^2-a_{kp}a_{1p}+a_{1p}a_{kp}a_{k+1p}^2 \leq 1-a_{k+1p}a_{1p}-a_{kp}a_{k+1p}+a_{1p}a_{kp}a_{k+1p}^2 \tag{5.13}$$

To prove (5.13) it suffices to prove the inequality

$$-a_{k+1p}^2-a_{kp}a_{1p} \leq -a_{k+1p}a_{1p}-a_{kp}a_{k+1p}$$

which is equivalent to proving the inequality:

$$a_{k+1p}^2+a_{kp}a_{1p}-a_{k+1p}a_{1p}-a_{kp}a_{k+1p} \geq 0$$

The left-hand side of the last inequality can be factorised:

$$a_{k+1p}^2+a_{kp}a_{1p}-a_{k+1p}a_{1p}-a_{kp}a_{k+1p} = \left(a_{k+1p}-a_{1p}\right)\left(a_{k+1p}-a_{kp}\right) \tag{5.14}$$

Since the values are arranged in ascending order ($a_{1p} \leq a_{2p} \leq \ldots \leq a_{kp} \leq a_{k+1p}$), in (5.14) $a_{k+1p}-a_{1p} \geq 0$ and $a_{k+1p}-a_{kp} \geq 0$. The product $(a_{k+1p}-a_{1p})(a_{k+1p}-a_{kp})$ in the right-hand side of (5.14) is therefore non-negative which proves inequality (5.12). Substituting $(1-a_{kp}a_{k+1p})(1-a_{k+1p}a_{1p})$ instead of $\left(1-a_{kp}a_{1p}\right)\left(1-a_{k+1p}^2\right)$ in the right-hand side of (5.11) will only strengthen the inequality and the inequality

$$\left(1-a_{1p}^2\right)\left(1-a_{2p}^2\right)\ldots\left(1-a_{kp}^2\right)\left(1-a_{k+1p}^2\right) \leq \left(1-a_{1p}a_{2p}\right)\left(1-a_{2p}a_{3p}\right)$$
$$\ldots\left(1-a_{kp}a_{k+1p}\right)\left(1-a_{k+1p}a_{1p}\right)$$
$$0 \leq a_{ip} \leq 1$$

is obtained. This proves the induction step for $k+1$ terms which, together with the trivial case for $n = 2$, completes the proof of inequality (5.8).

A natural reverse engineering of inequality (5.8) can now be given in terms of reliability of a series-parallel system including components that work and fail *independently* from one another. If the variables a_i in inequality (5.8) are interpreted as probabilities of failure of statistically independent components A_i, the left-hand side of inequality (5.8) gives the reliability of the system configuration in Figure 5.4a while the right-hand side of inequality (5.8) gives the reliability of

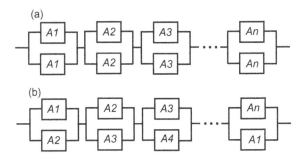

FIGURE 5.4 Reliability networks of two alternative systems built with the same types and number of components $A_1, A_2,...,A_n$: (a) with a symmetrical arrangement of the redundant components and (b) with an asymmetrical arrangement of the redundant components.

the system configuration in Figure 5.4b. Figure 5.4a and 5.4b depict reliability networks of common systems with interchangeable active redundancies at a component level.

Suppose that components A_i ($i = 1, ..., n$) stand for interchangeable sensors of n different types logically arranged in series. The sensors collect critical information from n zones in a system. For the system to operate successfully, at least a single sensor from each zone (block) must be operational. Each zone (block) includes a pair of sensors working in parallel.

Any particular type of sensor can work as a redundant sensor in any zone (block).

The reverse engineering of inequality (5.8) yields new knowledge: For systems built with components that work and fail independently from one another, the reliability of the system with an asymmetrical arrangement of the active redundancies in Figure 5.4b is always greater than the reliability of the system in Figure 5.4a with a symmetrical arrangement of the redundancies (Todinov, 2023a). This result holds *irrespective of the actual reliabilities of the components building the systems or their ranking*. A conclusion has been reached that the natural arrangement of the same type redundancies $(A_1/A_1, A_2/A_2,...,A_n/A_n)$ results in a smaller system reliability compared to an asymmetrical arrangement $A_1/A_2, A_2/A_3,...,A_n/A_1$.

To maximize the reliability of the original system, in the presence of a total uncertainty about the probabilities of failure of the individual components, the arrangement of the interchangeable redundancies must be such that the original symmetric arrangement is completely destroyed in all parts of the system.

These new insights, uncovered through the reverse engineering of the algebraic inequality 5.8, eluded system reliability experts for decades.

We must point out that no particular ranking of the components by their reliability is necessary for the method to be applied. To demonstrate this, we select the series-parallel system in Figure 5.4a consisting of 4 sections connected in series ($n = 4$).

The probabilities of failure of the components and the reliabilities of systems (5.4c) and (5.4c), for $n = 4$, are given in Table 5.1. The values x_1, x_2, x_3 and x_4

TABLE 5.1

Reliabilities of the Systems in Figure 5.4 ($n = 4$) for Different Values of the Probabilities of Failure of the Components Building the Systems

x_1	x_2	x_3	x_4	R_a	R_b
0.8	0.4	0.9	0.3	0.052	0.241
0.2	0.6	0.35	0.78	0.211	0.426
0.9	0.25	0.35	0.88	0.056	0.137
0.15	0.45	0.12	0.54	0.544	0.758
0.5	0.3	0.75	0.3	0.271	0.686
0.65	0.52	0.30	0.86	0.0998	0.183

correspond to the probabilities of failure of the components A_1, A_2, A_3 and A_4 in Figure 5.4. The values $R_a = \left(1 - x_1^2\right)\left(1 - x_2^2\right)\left(1 - x_3^2\right)\left(1 - x_4^2\right)$ are the reliabilities of the systems with symmetric redundancies while the values $R_b = (1 - x_1 x_2)(1 - x_2 x_3)$ $(1 - x_3 x_4)(1 - x_4 x_1)$ are the reliabilities of the systems with asymmetric redundancies.

As can be verified from Table 5.1, in all instances of ordering the component probabilities of failure, the asymmetric arrangement of redundancies always leads to a system with superior reliability ($R_b > R_a$). Not a single contradiction to this trend will be obtained if Table 5.1 is extended indefinitely.

The same approach to improving system reliability is valid for other interchangeable components, for example, for interchangeable switches of different types, seals, pumps, etc.

The knowledge derived from the reverse engineering of inequality (5.8) can be used for optimising series-parallel systems. The system reliability is substantially increased if the symmetry in the arrangement of the different types of redundant components is destroyed.

If the probabilities of failure a_i of the components are known, the reliability of the system in Figure 5.5a, with a symmetrical arrangement of the redundant components, can be maximised by arranging the components in the upper branches in ascending order of their probabilities of failure while the components in the lower

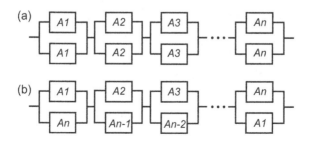

FIGURE 5.5   (a) Reliability network with interchangeable redundancies. (b) Reliability network of the system characterised by the largest system reliability.

branches are arranged in descending order of their probabilities of failure. The reliability network of the system with the largest reliability is given in Figure 5.5b.

Without loss of generality, we can always assume that the components A_1, $A_2,...,A_n$ in the upper branches in Figure 5.5b are arranged according to their probabilities of failure in ascending order. Suppose that the main components (in the upper branches of Figure 5.5b) are arranged in ascending order of their probabilities of failure: $a_1 \leq a_2 \leq ... \leq a_n$. The redundant components (the lower branches of Figure 5.5b) are arranged in descending order of their probabilities of failure: $a_n \geq a_{n-1} \geq ... \geq a_1$.

It can then be shown that the permutation in Figure 5.5b is characterised by the largest reliability, compared to the any other permutation (including the permutation in Figure 5.5a). Since the reliability of the system in Figure 5.5b is given by $R = (1 - a_1a_n)(1 - a_2a_{n-1})...(1 - a_na_1)$, it is effectively required to show that the inequality (Todinov, 2023a)

$$\left(1-a_1a_n\right)\left(1-a_2a_{n-1}\right)...\left(1-a_na_1\right) \geq \left(1-a_1a_{p1}\right)\left(1-a_2a_{p2}\right)...\left(1-a_na_{pn}\right)$$
$$0 \leq \left(1-a_ia_j\right) \leq 1$$
(5.15)

holds, where a_{p1}, a_{p2}, ..., a_{pn} is any particular permutation of the components (probabilities of failure) in the lower branches.

Proof: Inequality (5.15) can be proved by using *the extreme principle*. Suppose that there is an arrangement where the components in the lower branches are not arranged in descending order of their probabilities of failure and the system reliability given by the product $(1 - a_1a_{p1})(1 - a_2a_{p2})...(1 - a_na_{pn})$ is the largest possible. In this case, there must exist at least two terms $(1 - a_ia_{px})(1 - a_ja_{py})$ where $i < j$; $a_i < a_j$; and $a_{px} < a_{py}$. Otherwise, if no such terms can be found, for which $a_{px} < a_{py}$, the components in the lower branches would have been arranged in descending order.

We will show that if $a_{px} < a_{py}$, the system reliability given by the product $(1 - a_1a_{p1})(1 - a_2a_{p2})...(1 - a_na_{pn})$ cannot be the largest possible which leads to a contradiction with the assumption that this is the largest possible system reliability.

Compare the product $(1 - a_ia_{px})(1 - a_ja_{py})$ with the product $(1 - a_ia_{py})(1 - a_ja_{px})$ obtained by swapping the redundant components with indices 'px' and 'py' in the lower branches. We will show that

$$\left(1-a_ia_{px}\right)\left(1-a_ja_{py}\right) < \left(1-a_ia_{py}\right)\left(1-a_ja_{px}\right)$$
(5.16)

Expanding the left- and right-hand side of inequality (5.16) leads to the equivalent inequality

$$-a_ia_{px} - a_ja_{py} < -a_ia_{py} - a_ja_{px}$$
(5.17)

Inequality (5.17) is equivalent to the inequality

$$\left(a_{py} - a_{px}\right)\left(a_j - a_i\right) > 0 \tag{5.18}$$

Inequality (5.18) is true because $a_j > a_i$ and $a_{py} > a_{px}$. This shows that inequality (5.16) holds and, contrary to the assumption that the system has the largest possible reliability, the system reliability is not the largest possible. This means that a larger product

$$\left(1 - a_1 a_{p1}\right)\left(1 - a_2 a_{p2}\right)...\left(1 - a_n a_{pn}\right)$$

cannot be obtained from a permutation $a_{p1}, a_{p2}, ..., a_{pn}$ that is different from the permutation corresponding to arranging the values a_{pi} in descending order: $a_{p1} \geq a_{p2} \geq, ..., \geq a_{pn}$ which completes the proof of inequality (5.15).

It is difficult to see how these general results related to the reliability of a series-parallel system could be obtained without proving the non-trivial inequalities (5.8) and (5.15).

The difference in the system reliabilities of the competing systems in Figure 5.4a, 5.4b and Figure 5.5a, 5.5b, for example, can be very large as the next, deliberately simplified example based on $n = 4$ components, demonstrates. Thus, for interchangeable sensors of types A, B, C and D, characterised by probabilities of failure of $a_1 = 0.28$, $a_2 = 0.53$, $a_3 = 0.82$ and $a_4 = 0.85$), for two years of continuous operation, the arrangement in Figure 5.4a ($n = 4$) is characterised by a system reliability of

$$R_a = \left(1 - 0.28^2\right)\left(1 - 0.53^2\right)\left(1 - 0.82^2\right)\left(1 - 0.85^2\right) = 0.06$$

while the arrangement in Figure 5.4b is characterised by a system reliability of

$$R_b = \left(1 - 0.28 \times 0.53\right)\left(1 - 0.53 \times 0.82\right)\left(1 - 0.82 \times 0.85\right)\left(1 - 0.85 \times 0.28\right) = 0.11$$

The largest system reliability is obtained for the arrangement in Figure 5.5b ($n = 4$):

$$R_b = \left(1 - 0.28 \times 0.85\right)\left(1 - 0.53 \times 0.82\right)\left(1 - 0.82 \times 0.53\right)\left(1 - 0.85 \times 0.28\right) = 0.186$$

An inequality similar to inequality (5.8) can also be proved for a more complex system, for example, for the system in Figure 5.6a.

It can be shown that the inequality

$$\left(1 - a_1^m\right)\left(1 - a_2^m\right)...\left(1 - a_n^m\right) \leq \left(1 - a_1^{m-t} a_2^t\right)\left(1 - a_2^{m-t} a_3^t\right)...\left(1 - a_n^{m-t} a_1^t\right) \tag{5.19}$$

holds, where $m > 2$ and $1 \leq t \leq m$.

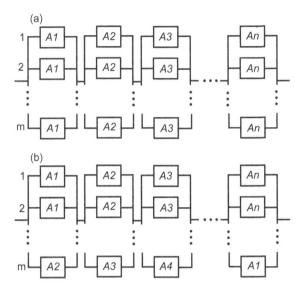

FIGURE 5.6 Reliability networks of two alternative systems: (a) a system with $m - 1$ active redundant components and (b) a system for which the redundant components in one of the branches have been cyclically shifted.

As a result, the system in Figure 5.6a, has a lower reliability compared to the system with asymmetric redundancy arrangement (corresponding to $t = 1$) in Figure 5.6b.

In this case, the inequality

$$\left(1-a_1^m\right)\left(1-a_2^m\right)...\left(1-a_n^m\right) \leq \left(1-a_1^{m-1}a_2\right)\left(1-a_2^{m-1}a_3\right)...\left(1-a_n^{m-1}a_1\right) \quad (5.20)$$

is obtained from (5.19), which corresponds to $t = 1$. The proof of inequality (5.19) is similar to the proof presented for inequality (5.8) and has been given in the Appendix.

5.3 PRACTICAL APPLICATIONS

The result related to maximising system reliability can, for example, be used for any system involving a new (N), medium-age (M) and old component (O) and interchangeable redundancies (Figure 5.7a). Let the probability of failure of the new, medium age and old component be a_N, a_M, and a_o, respectively. Making the natural assumption $a_N < a_M < a_o$ for the probabilities of failure of the components, arranging the components and the redundancies as is shown in Figure 5.7b yields the largest system reliability. This arrangement always brings the largest system reliability, irrespective of the specific probabilities of failure characterising the components. The only requirement is the ranking assumption $a_N < a_M < a_O$ related

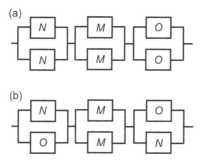

FIGURE 5.7 a) Reliability network with interchangeable redundancies involving new ('N'), medium-age ('M') and old components ('O') b) The reliability network of the system characterised by the largest reliability.

to the probabilities of failure which, after eliminating early-life failures, commonly holds for new, medium-age and old components.

More importantly, the proposed approach can be utilized in diverse systems that involve interchangeable redundant components for which the ranking of their probabilities of failure is unknown.

Interchangeable redundancies can involve different types of components, which correspond to real-world applications. For example, interchangeable redundancies of different types might include interchangeable seals of various kinds, such as O-ring seals, cartridge seals, pressurized and unpressurized seals, etc. Interchangeable sensors could measure the same quantity but operate on different physical principles. For instance, the temperature in two zones could be measured with thermocouples of different types, thermal resistors, or bi-metal thermometers. Similarly, the pressure in two zones could be measured by bellows-type pressure gauges, manometer pressure gauges, piezometer pressure gauges, Bourdon tube pressure gauges, and capsule pressure gauges.

Interchangeable redundancies can also involve components of the same type but of different varieties. Examples of components of the same type but different varieties include components of a particular type but of different ages. Additionally, components of a particular type can vary if they are produced by different manufacturers. Components of the same type but different varieties also differ in their reliability values. Typically, new components of a particular type have greater reliability than older components of the same type. Similarly, components of the same type produced by different manufacturers also differ in their reliability.

Consider a scenario where the temperature, pressure, concentration, or fluid level within each of two distinct zones, A and B (as illustrated in Figure 5.8a and 5.8b), must be measured and relayed to a control device (C). The control device then compares the readings obtained from both zones. If the disparity between the measurements in zones A and B surpasses a critical threshold, the control device activates a safety system to mitigate potential risks.

Within each zone, the sensors are logically arranged in parallel. To enable a valid comparison, signals from at least one sensor in zone A and zone B must be received by the control device C.

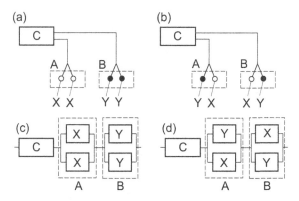

FIGURE 5.8 (a and b) Application example involving redundant sensors of type X and type Y; (c and d) Reliability networks corresponding to the systems in Figure 5.8a and b.

Suppose that the sensors are of two types: X and Y. The reliability networks of the systems in Figure 5.8a and b are provided in Figures 5.8c and 5.8d, respectively.

If $0 \le c \le 1$, $0 \le x \le 1$ and $0 \le y \le 1$ stand for the probabilities of failure of components C, X and Y, the reliability of the system configuration in Figure 5.8a is given by $R_a = (1 - c)(1 - x^2)(1 - y^2)$ while the reliability of the system configuration in Figure 5.8b is given by $R_b = (1 - c)(1 - xy)(1 - xy)$. The reverse engineering of the correct algebraic inequality

$$R_a = \left(1-c\right)\left(1-x^2\right)\left(1-y^2\right) \le R_b = \left(1-c\right)\left(1-xy\right)\left(1-xy\right)$$

yields that the reliability of the system in Figure 5.8b consistently surpasses that of the system in Figure 5.8a. The last inequality holds true regardless of the individual probabilities of failure x and y of sensors type X and type Y, as well as the probability of failure c of the control device (Todinov, 2023b).

The difference in the system reliabilities of the alternatives in Figure 5.8a and b can be significant and this can be demonstrated with the specific probabilities of failure: $c = 0.1$, $x = 0.17$ and $y = 0.86$ of components C,X and Y. The reliabilities of the alternative systems in Figure 5.8a and b are:

$$R_a = \left(1-c\right)\left(1-x^2\right)\left(1-y^2\right) = \left(1-0.1\right)\left(1-0.17^2\right)\left(1-0.86^2\right) = 0.23$$

and

$$R_b = \left(1-c\right)\left(1-xy\right)\left(1-xy\right) = \left(1-0.1\right)\left(1-0.17\times0.86\right)^2 = 0.66$$

As a result, the level of system reliability achieved through an asymmetric configuration of the sensors is 2.9 times greater than that of a symmetric arrangement.

In conclusion, it is not beneficial with respect to system reliability to place sensors of the same type in the same zone. Instead, mixing sensor types is recommended to achieve an asymmetrical configuration of redundancies and a significant enhancement of the overall system reliability.

The next application example (Todinov, 2023b) comprises four parallel pipelines that transport toxic fluid (Figure 5.9a). Each pipeline incorporates flanges sealed by two O-ring seals, with the second O-ring serving as a redundant seal. It's crucial to note that a solitary functional seal in a flange is sufficient to isolate the toxic fluid from the surrounding environment.

The O-ring seals utilized in this system are of two different materials: material X (represented by open circles in Figure 5.9) and material Y (represented by filled circles in Figure 5.9). For the system of four pipelines with flanges to operate reliably, there must be no leak of toxic fluid from any of the flanges.

The flanges are logically arranged in series because all of the flanges must isolate the toxic fluid in order to prevent its release in the environment. On the other hand, the seals within each flange are arranged in parallel because the reliable operation of just one seal is adequate to prevent a leak from the flange. As a result, the corresponding reliability networks for the systems illustrated in Figure 5.9a and 5.9b can be found in Figure 5.9c and 5.9d, respectively.

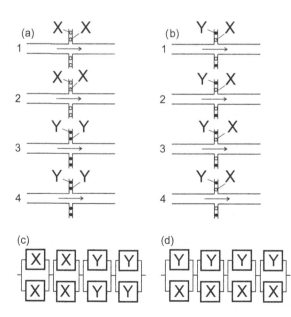

FIGURE 5.9 Application example involving four pipelines and flanges with seals from two different materials X and Y; (a) the original system with a symmetric arrangement of the seals; (b) the system with superior reliability with an asymmetric arrangement of the seals, and (c) and (d) reliability networks corresponding to the systems in Figure 5.8a and b.

Let 'x' and 'y' denote the probability of failure of seals made from material X and Y, respectively. The reliabilities of the systems of seals depicted in Figure 5.8a and b are then given by:

$$R_a = \left(1-x^2\right)\left(1-x^2\right)\left(1-y^2\right)\left(1-y^2\right)$$

and

$$R_b = \left(1-xy\right)\left(1-xy\right)\left(1-xy\right)\left(1-xy\right)$$

The reverse engineering of the correct algebraic inequality:

$$R_a = \left(1-x^2\right)\left(1-x^2\right)\left(1-y^2\right)\left(1-y^2\right) \le R_b = \left(1-xy\right)\left(1-xy\right)\left(1-xy\right)\left(1-xy\right)$$

yields that the seals arrangement in Figure 5.9b is the more reliable seals arrangement, irrespective of the actual reliabilities of the seals from material X and Y. This algebraic inequality can be proved by using the simpler auxiliary inequality $(1 - x^2)(1 - y^2) \le (1 - xy)(1 - xy)$ which is proved easily by expanding the brackets. *The conclusion is that mixing the seals from both types in each flange yields a more reliable system.* This conclusion has been reached in total absence of knowledge about the probabilities of failure of the seals or their ranking.

The system reliability enhancement can be dramatic, as the next numerical example demonstrates. Thus, for probabilities of failure $x = 0.17$, $y = 0.86$ related to the seals of type X and type Y, correspondingly, the reliability of the system configuration in Figure 5.9a is:

$$R_a = \left(1-x^2\right)^2\left(1-y^2\right)^2 = \left(1-0.17^2\right)^2\left(1-0.86^2\right)^2 = 0.064$$

while the reliability of the system configuration in Figure 5.9b is

$$R_b = \left(1-xy\right)^2\left(1-yx\right)^2 = \left(1-0.17\times0.86\right)^2\left(1-0.86\times0.17\right)^2 = 0.53$$

The system reliability achieved through the asymmetric configuration of redundancies is 8.3 times greater than that of a symmetric arrangement!

APPENDIX 5.A: PROOF OF INEQUALITY (5.19)

Inequality (5.19) can be proved by induction, by proving first the case for two sections ($n = 2$):

$$\left(1-a_1^m\right)\left(1-a_2^m\right) \le \left(1-a_1^{m-t}a_2^t\right)\left(1-a_2^{m-t}a_1^t\right) \tag{5.A.1}$$

$0 \le a_1 \le 1; 0 \le a_2 \le 1; a_1 \le a_2, 1 \le t \le m$. Proving this inequality is equivalent to proving the equivalent inequality

$$1 - a_1^m - a_2^m + a_1^m a_2^m \le 1 - a_1^{m-t} a_2^t - a_2^{m-t} a_1^t + a_1^m a_2^m \qquad (5.A.2)$$

which, in turn, can be proved by proving the equivalent inequality

$$a_1^m + a_2^m - a_1^{m-t} a_2^t - a_1^t a_2^{m-t} \ge 0 \qquad (5.A.3)$$

Inequality (5.A.3) however, is true because $a_1^m + a_2^m - a_1^{m-t} a_2^t - a_1^t a_2^{m-t} = \left(a_2^t - a_1^t \right) \left(a_2^{m-t} - a_1^{m-t} \right)$ is non-negative.

Inequality (5.19) can now be proved by induction. For $n = 2$, inequality (5.19) coincides with inequality (5.A.2) which has been shown to be true.

The probabilities of failure of the components can always be ordered in ascending $(a_1 \le a_2 \le \ldots \le a_k \le a_{k+1})$ order. Let us assume that inequality (5.19) is true for $n = k$ (induction hypothesis):

$$\left(1 - a_1^m \right)\left(1 - a_2^m \right)\ldots\left(1 - a_k^m \right) \le \left(1 - a_1^{m-t} a_2^t \right)\left(1 - a_2^{m-t} a_3^t \right)\ldots\left(1 - a_k^{m-t} a_1^t \right) \quad (5.A.4)$$

We will show that the inequality is also valid for $n = k + 1$.

Multiplying both sides of inequality (5.A.4) by $\left(1 - a_{k+1}^m \right)$ gives the inequality

$$\left(1 - a_1^m \right)\left(1 - a_2^m \right)\ldots\left(1 - a_k^m \right)\left(1 - a_{k+1}^m \right) \le \left(1 - a_1^{m-t} a_2^t \right)\left(1 - a_2^{m-t} a_3^t \right)$$
$$\ldots\left(1 - a_k^{m-t} a_1^t \right)\left(1 - a_{k+1}^m \right) \qquad (5.A.5)$$

If it can be shown that

$$\left(1 - a_k^{m-t} a_1^t \right)\left(1 - a_{k+1}^m \right) \le \left(1 - a_k^{m-t} a_{k+1}^t \right)\left(1 - a_{k+1}^{m-t} a_1^t \right) \qquad (5.A.6)$$

This means that replacing the expression $\left(1 - a_k^{m-t} a_1^t \right)\left(1 - a_{k+1}^m \right)$ in the right-hand side of inequality (5.A.5) by the larger expression $\left(1 - a_k^{m-t} a_{k+1}^t \right)\left(1 - a_{k+1}^{m-t} a_1^t \right)$, will only strengthen inequality (5.A.5).

To prove inequality (5.A.6), the equivalent inequality:

$$1 - a_{k+1}^m - a_k^{m-t} a_1^t + a_1^t a_k^{m-t} a_{k+1}^m \le 1 - a_{k+1}^{m-t} a_1^t - a_k^{m-t} a_{k+1}^t + a_1^t a_k^{m-t} a_{k+1}^m \quad (5.A.7)$$

must be proved, which is obtained from expanding the left- and right-hand side of (5.A.6). Proving inequality (5.A.7) is equivalent to proving

$$a_{k+1}^m + a_k^{m-t} a_1^t - a_{k+1}^{m-t} a_1^t - a_k^{m-t} a_{k+1}^t \ge 0 \qquad (5.A.8)$$

The left-hand side of (5.A.8) can be factorised as:

$$a_{k+1}^m + a_k^{m-t}a_1^t - a_{k+1}^{m-t}a_1^t - a_k^{m-t}a_{k+1}^t = \left(a_{k+1}^t - a_1^t\right)\left(a_{k+1}^{m-t} - a_k^{m-t}\right) \qquad (5.A.9)$$

and because $a_1 \le a_2 \le \ldots \le a_k \le a_{k+1}$, it follows that $\left(a_{k+1}^t - a_1^t\right)\left(a_{k+1}^{m-t} - a_k^{m-t}\right) \ge 0$.
The case $n = k + 1$ has been proved. Because inequality (5.19) is true for $n = 2$, the inequality is also true for $n = 3$, $n = 4$ and for any other $n \ge 2$.

6 Enhancing the Reliability of Series-Parallel Systems with Multiple Redundancies by Reverse Engineering of Algebraic Inequalities

This chapter extends the previous chapter by reverse engineering of new algebraic inequalities that are applicable to series-parallel systems with multiple redundancies. The reliability of such systems is enhanced by using reverse engineering of correct algebraic inequalities.

The reverse engineering of correct algebraic inequalities often leads to a projection of a new physical reality characterised by a distinct signature: the algebraic inequality itself. Different physical systems or processes may have a single algebraic inequality as a signature. In this respect, these physical systems and processes form many-to-one mapping with the inequality and new properties/ behaviour related to diverse physical systems can be inferred from the same abstract inequality.

Furthermore, the proposed approach for enhancing system reliability is domain-independent and does not require any information about the reliabilities of the components or their ranking. To prove the proposed system reliability inequalities, a novel method based on segmentation and permutation of the variable values has been developed.

6.1   IMPROVING THE RELIABILITY OF SERIES-PARALLEL SYSTEMS IN CASES OF UNKNOWN RELIABILITIES OF THE COMPONENTS AND SYMMETRIC ARRANGEMENT OF THE REDUNDANCIES

Reverse engineering of algebraic inequalities can be used to improve the reliability of series-parallel systems in case of unknown reliabilities of the components or their ranking. It is assumed that each component in the systems has the same number of interchangeable redundancies. Consider the correct abstract algebraic inequality:

$$\left(1-x^3\right)\left(1-y^3\right)\left(1-z^3\right) \leq \left(1-x^2y\right)\left(1-y^2z\right)\left(1-z^2x\right) \tag{6.1}$$

which will be proved rigorously in Section 6.4. In inequality (6.1): $0 \leq x \leq 1, 0 \leq y \leq 1, 0 \leq z \leq 1$.

Let the variables x, y and z in inequality (6.1) be physically interpreted as 'probabilities of failure' of components of type X, Y and Z. It can be shown that, in this case, the left- and the right-hand side of inequality (6.1) can be physically interpreted as reliabilities of two alternative series-parallel systems.

Indeed, consider the physical system in Figure 6.1 (Todinov, 2023b). It features three pipelines with three valves of types X, Y and Z, physically arranged in series. All valves are initially open. With respect to stopping the fluid in all three pipelines, the reliability networks corresponding to the physical arrangements (a) and (b) are given by Figure 6.2a and 6.2b, correspondingly. These reliability networks

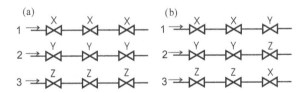

FIGURE 6.1   Functional diagrams of two different arrangements of valves on three pipelines.

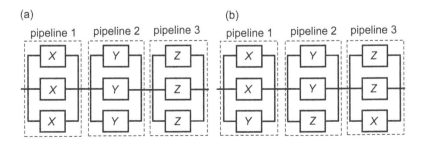

FIGURE 6.2   Reliability networks of the systems in Figure 6.1.

represent the logical arrangement of the valves, which is different from their physical arrangement. Each section of three valves in parallel in Figure 6.2a or Figure 6.2b corresponds to the three valves in series on each of the pipelines in Figure 6.1a or Figure 6.1b.

If the probabilities of failures of the valves of types X, Y and Z are denoted by x, y and z, respectively ($0 \le x \le 1$, $0 \le y \le 1$, $0 \le z \le 1$), the reliability R_a of the system in Figure 6.2a is given by $R_a = (1 - x^3)(1 - y^3)(1 - z^3)$ while the reliability R_b of the system in Figure 6.2b is given by $R_b = (1 - x^2y)(1 - y^2z)(1 - z^2x)$. The probabilities of failure x, y and z, of the valves are unknown.

The reverse engineering of inequality (6.1) yields that the left-hand side of inequality (6.1) corresponds to the reliability of the system in Figure 6.2a while the right-hand side of the inequality corresponds to the reliability of the system in Figure 6.2b.

According to inequality (6.1), the reliability of the system in Figure 6.2b is superior to the reliability of the system in Figure 6.1a and this conclusion has been made *in total absence of knowledge regarding the probabilities of failure x, y and z of the three different types of valves* or their ranking (Todinov, 2023b).

Inequalities similar to inequality (6.1) can be reverse engineered relatively easily because products of the type $\left(1 - x_1^m\right)\left(1 - x_2^m\right)...\left(1 - x_n^m\right)$ where x_i is the probability of failure of component i can be interpreted directly as reliability of series-parallel systems with m components in parallel in each section in series.

In this connection, inequalities (6.2)–(6.5), where $0 \le x, y, z \le 1$, can be proved rigorously and reverse-engineered as reliabilities of series-parallel systems.

$$\left(1-x^2\right)\left(1-y^2\right)\left(1-z^2\right) \le \left(1-xy\right)\left(1-yz\right)\left(1-zx\right) \tag{6.2}$$

$$\left(1-x^3\right)\left(1-y^3\right)\left(1-z^3\right) \le \left(1-xyz\right)\left(1-yzx\right)\left(1-zxy\right) \tag{6.3}$$

$$\left(1-x^4\right)\left(1-y^4\right)\left(1-z^4\right) \le \left(1-x^2y^2\right)\left(1-y^2z^2\right)\left(1-z^2x^2\right) \tag{6.4}$$

$$\left(1-x^4\right)\left(1-y^4\right)\left(1-z^4\right) \le \left(1-x^2yz\right)\left(1-y^2zx\right)\left(1-z^2xy\right) \tag{6.5}$$

If x, y and z are interpreted as probabilities of failure of components from types X, Y and Z, the reverse engineering of inequalities (6.2)–(6.5) yields that their left- and right-hand sides correspond to the reliabilities of the series-parallel systems in Figures 6.3–6.6, correspondingly. The left hand-sides of the inequalities correspond to the systems 'a' while the right-hand sides correspond to the reliability of systems 'b'.

The reverse engineering of inequalities (6.1)–(6.5) yields new knowledge in system reliability. The system reliabilities of the asymmetric redundancy arrangements 'b' are *always* greater than the system reliabilities of the symmetric redundancy arrangements 'a'. As a result, inequalities (6.1)–(6.5) permit enhancing the

(a)   (b)

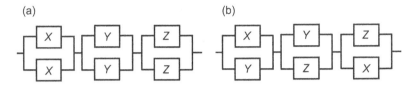

FIGURE 6.3   (a) A system obtained from the physical interpretation of the left-hand side of inequality (6.2) and (b) a system obtained from the physical interpretation of the right-hand side of inequality (6.2).

(a)   (b)

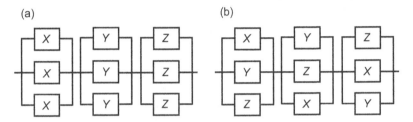

FIGURE 6.4   (a) A system obtained from the physical interpretation of the left-hand side of inequality (6.3) and (b) a system obtained from the physical interpretation of the right-hand side of inequality (6.3).

(a)   (b)

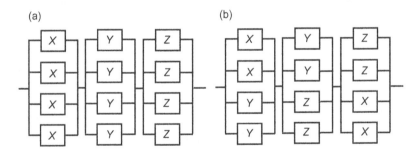

FIGURE 6.5   (a) A system obtained from the physical interpretation of the left-hand side of inequality (6.4) and (b) a system obtained from the physical interpretation of the right-hand side of inequality (6.4).

reliability of series-parallel systems with the same number multiple interchangeable redundant components, in total absence of knowledge related to the reliabilities of the components building the systems or their ranking (Todinov, 2023b).

The system reliability enhancement can be dramatic, as the next numerical example demonstrates. Thus, for probabilities of failure $x = 0.08$, $y = 0.23$, $z = 0.91$ associated with a given period of operation, the reliability of the system configuration in Figure 6.2a is:

$$R_a = \left(1-x^3\right)\left(1-y^3\right)\left(1-z^3\right) = \left(1-0.08^3\right)\left(1-0.23^3\right)\left(1-0.91^3\right) = 0.24$$

(a)   (b)

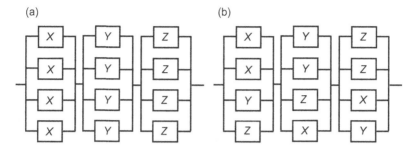

FIGURE 6.6   (a) A system obtained from the physical interpretation of inequality (6.5) and (b) a system obtained from the physical interpretation of the right-hand side of inequality (6.5).

while the reliability of the system configuration in Figure 6.2b is

$$R_b = \left(1 - x^2 y\right)\left(1 - y^2 z\right)\left(1 - z^2 x\right)$$
$$= \left(1 - 0.08^2 \times 0.23\right)\left(1 - 0.23^2 \times 0.91\right)\left(1 - 0.91^2 \times 0.08\right) = 0.89$$

The system reliability achieved through the asymmetric configuration of redundancies is 3.7 times greater than that of a symmetric arrangement!

These analyses indicate that in systems featuring the same number of multiple interchangeable redundant components, a symmetric configuration of the redundant components consistently results in lower system reliability compared to an asymmetric arrangement. This conclusion remains valid irrespective of the failure probabilities associated with components or their ranking.

Note that the results obtained for systems with an equal number of redundancies cannot automatically be extrapolated to systems with a different number of redundancies in each section in series because systems with a different number of redundancies in each section are no longer symmetrical.

6.2   IMPROVING THE RELIABILITY OF SERIES-PARALLEL SYSTEMS IN CASE OF UNKNOWN RELIABILITIES OF THE COMPONENTS AND ASYMMETRIC ARRANGEMENT OF THE REDUNDANCIES

Consider now the algebraic inequality (Todinov, 2023b):

$$\left(1 - x^2 y\right)\left(1 - y^2 z\right)\left(1 - z^2 x\right) \le \left(1 - xyz\right)\left(1 - xyz\right)\left(1 - xyz\right) \qquad (6.6)$$

which will be proved rigorously in Section 6.3.2.

The reverse engineering of this inequality consists of a physical interpretation of its left-hand side as the reliability of the system in Figure 6.7a while the physical interpretation of its right-hand side yields the reliability of the system in Figure 6.7b.

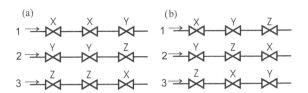

FIGURE 6.7 Functional diagrams of two different arrangements of valves on three pipelines. Each pipeline includes valves of different type.

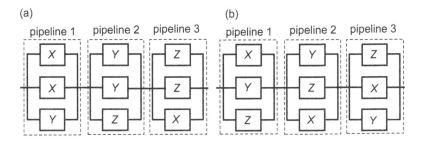

FIGURE 6.8 Reliability networks of the systems in Figure 6.7.

The physical system in Figure 6.7a features three pipelines with three valves of types X, Y and Z, physically arranged in series. All valves are initially open. With respect to stopping the fluid in all three pipelines, the reliability networks corresponding to the asymmetric physical arrangements (a) and (b) are given by Figure 6.8a and b, correspondingly. Each section of three valves in parallel in Figure 6.8 corresponds to the three valves in series on each of the pipelines in Figure 6.7. If the probabilities of failure of types X, Y and Z valves are denoted by x, y and z, respectively, the reliability of the system in Figure 6.8a is given by $R_a = (1 - x^2 y)(1 - y^2 z)(1 - z^2 x)$ while the reliability of the system in Figure 6.8b is given by $R_b = (1 - xyz)(1 - xyz)(1 - xyz)$.

According to the physical interpretation of inequality (6.6), the reliability of the system in Figure 6.7b is superior to the reliability of the system in Figure 6.7a and this conclusion has been made in total absence of knowledge regarding the probabilities of failure x, y and z of the valves or their ranking.

The following inequalities can be proved rigorously and reverse engineered as reliabilities of physical systems (Todinov, 2023b):

$$\left(1-x^3 y\right)\left(1-y^3 z\right)\left(1-z^3 x\right) \le \left(1-x^2 yz\right)\left(1-y^2 zx\right)\left(1-z^2 xy\right) \tag{6.7}$$

$$\left(1-x^2 y^2\right)\left(1-y^2 z^2\right)\left(1-z^2 x^2\right) \le \left(1-x^2 yz\right)\left(1-y^2 zx\right)\left(1-z^2 xy\right) \tag{6.8}$$

$$\begin{aligned}
\left(1-x^4 y\right)\left(1-y^4 z\right)&\left(1-z^4 u\right)\left(1-u^4 x\right) \\
&\le \left(1-x^2 yzu\right)\left(1-xy^2 zu\right)\left(1-xyz^2 u\right)\left(1-xyzu^2\right)
\end{aligned} \tag{6.9}$$

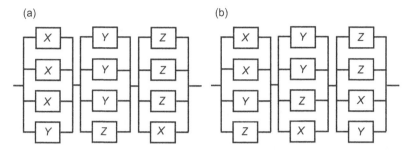

FIGURE 6.9 (a) A system obtained from the physical interpretation of the left-hand side of inequality (6.7) and (b) a system obtained from the physical interpretation of the right-hand side of inequality (6.7).

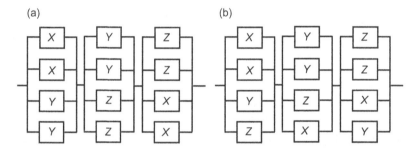

FIGURE 6.10 (a) A system obtained from the physical interpretation of the left-hand side of inequality (6.8) and (b) a system obtained from the physical interpretation of the right-hand side of inequality (6.8).

where $0 \leq x, y, z, u \leq 1$.

If x, y, z, u are physically interpreted as probabilities of failure of components X, Y, Z and U, the left- and right-hand sides of inequalities (6.6)–(6.9) correspond to the reliabilities of the series-parallel systems in Figures 6.8, 6.9, 6.10 and 6.11, correspondingly.

The left hand-sides of the inequalities correspond to the reliabilities of the systems 'a' while the right-hand sides correspond to the reliabilities of systems 'b'.

The reverse engineering of the correct algebraic inequalities (6.6)–(6.9) yields that the system reliability of the asymmetric arrangements of redundant components in systems 'b' is *always* greater than the system reliability of the asymmetric arrangements of the components in systems 'a', *irrespective of the probabilities of failure of the separate components*. As a result, the reverse engineering of these inequalities permits also enhancing the reliability of series-parallel systems with asymmetric arrangement of the redundancies irrespective of the individual reliabilities of the components.

(a)   (b)

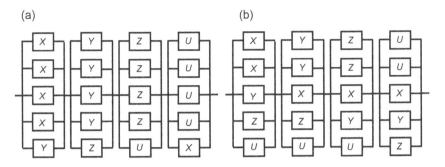

FIGURE 6.11 (a) A system obtained from the physical interpretation of the left-hand side of inequality (6.9) and (b) a system obtained from the physical interpretation of the right-hand side of inequality (6.9).

The system reliability enhancement can be dramatic, as the next numerical example demonstrates. Thus, for probabilities of failure $x = 0.17$, $y = 0.77$, $z = 0.86$ associated with a given period of operation, the reliability of the system configuration in Figure 6.8a is:

$$R_a = \left(1 - x^2 y\right)\left(1 - y^2 z\right)\left(1 - z^2 x\right)$$
$$= \left(1 - 0.17^2 \times 0.77\right)\left(1 - 0.77^2 \times 0.86\right)\left(1 - 0.86^2 \times 0.17\right) = 0.42$$

while the reliability of the system configuration in Figure 6.8b is

$$R_b = \left(1 - xyz\right)\left(1 - xyz\right)\left(1 - xyz\right) = \left(1 - 0.17 \times 0.77 \times 0.86\right)^3 = 0.7$$

The repositioning of redundancies has successfully improved system reliability by a factor of 1.67!

6.3 A NEW TECHNIQUE FOR PROVING SYSTEM RELIABILITY INEQUALITIES BASED ON SEGMENTATION AND PERMUTATION OF VARIABLES

The system reliability inequalities (6.1)–(6.9) can be proved by a novel technique based on segmentation and variable permutation. This technique will be illustrated by proving inequalities (6.1) and (6.6).

6.3.1 PROOF OF INEQUALITY (6.1)

The proof of the complex inequality (6.1) starts with proving the simpler inequalities

$$\left(1 - x^3\right)\left(1 - y^3\right) \leq \left(1 - x^2 y\right)\left(1 - y^2 x\right) \tag{6.10}$$

$$\left(1 - y^3\right)\left(1 - z^3\right) \leq \left(1 - y^2 z\right)\left(1 - z^2 y\right) \tag{6.11}$$

$$\left(1-z^{3}\right)\left(1-x^{3}\right)\leq\left(1-z^{2}x\right)\left(1-x^{2}z\right) \tag{6.12}$$

These simpler inequalities can be viewed as building segments whose multiplication leads to an inequality closely related to inequality (6.1).

Inequalities (6.10)–(6.12) are easily proved by direct manipulation. For example, proving inequality (6.10) is equivalent to proving the inequality $x^3 + y^3 \geq x^2y + y^2x$, which is equivalent to proving the inequality $x^3 - x^2y + y^3 - y^2x \geq 0$ or $x^2(x - y) - y^2(x - y) \geq 0$. The last inequality is equivalent to $(x - y)^2(x + y) \geq 0$ which is always true considering that x and y are non-negative. Because x, y and z are non-negative numbers between 0 and 1, the left- and right-hand sides of inequalities (6.10)–(6.12) are all non-negative. Multiplying the left- and right-hand sides of inequalities (6.10)–(6.12), therefore, will not alter the direction of the resultant inequality. Consequently, multiplying the left- and right-hand parts of inequalities (6.10)–(6.12) results in a correct inequality with the same direction:

$$\left[\left(1-x^{3}\right)\left(1-y^{3}\right)\left(1-z^{2}\right)\right]^{2}\leq\left(1-x^{2}y\right)\left(1-y^{2}z\right)\left(1-z^{2}x\right)$$
$$\left(1-y^{2}x\right)\left(1-z^{2}y\right)\left(1-x^{2}z\right) \tag{6.13}$$

that holds for any x, y and z for which $0 \leq x, y, z \leq 1$.

However, inequality (6.13) can only be true if and only if both inequalities (6.14) and (6.15) hold true simultaneously.

$$\left(1-x^{3}\right)\left(1-y^{3}\right)\left(1-z^{3}\right)\leq\left(1-x^{2}y\right)\left(1-y^{2}z\right)\left(1-z^{2}x\right) \tag{6.14}$$

$$\left(1-x^{3}\right)\left(1-y^{3}\right)\left(1-z^{3}\right)\leq\left(1-y^{2}x\right)\left(1-z^{2}y\right)\left(1-x^{2}z\right) \tag{6.15}$$

Note that the direction of both inequalities (6.14) and (6.15) cannot be '$\geq$' because, in this case, the product of the left- and right-hand sides of inequalities (6.14) and (6.15) would generate an inequality with a direction opposite to that of inequality (6.13).

Suppose that only the direction of inequality (6.15) is reversed. Hence, according to our assumption, the inequality

$$\left(1-x^{3}\right)\left(1-y^{3}\right)\left(1-z^{3}\right)\geq\left(1-y^{2}x\right)\left(1-z^{2}y\right)\left(1-x^{2}z\right) \tag{6.16}$$

holds for any x, y and z for which $0 \leq x, y, z \leq 1$. We aim to demonstrate that this particular assumption leads to a contradiction. To do so, we give the specific values $x = a$, $y = b$, $z = c$ for the variables x, y, z and from inequality (6.14) we get:

$$\left(1-a^{3}\right)\left(1-b^{3}\right)\left(1-c^{3}\right)\leq\left(1-a^{2}b\right)\left(1-b^{2}c\right)\left(1-c^{2}a\right) \tag{6.17}$$

Now let us interchange the values of variables x and y: $x = b$, $y = a$, $z = c$. From inequality (6.16) we get

$$\left(1-b^3\right)\left(1-a^3\right)\left(1-c^3\right) \geq \left(1-a^2b\right)\left(1-c^2a\right)\left(1-b^2c\right) \qquad (6.18)$$

The left- and right-hand sides of inequalities (6.17) and (6.18) are identical; however, the direction of inequality (6.18) is opposite to that of inequality (6.17). This contradiction is a result of the assumption made about the direction of inequality (6.16). Consequently, the direction of inequality (6.16) is the same as the direction of inequality (6.15).

In order to avoid repetition, we will refrain from reiterating the analogous arguments employed in the proof of inequalities (6.2) through (6.5). Additionally, these inequalities can be naturally extended to more than three variables.

6.3.2 PROOF OF INEQUALITY (6.6)

The poof of the complex inequality (6.6) starts with proving the simpler inequalities

$$\left(1-x^2y\right)\left(1-z^2y\right) \leq \left(1-xyz\right)^2 \qquad (6.19)$$

$$\left(1-y^2z\right)\left(1-x^2z\right) \leq \left(1-xyz\right)^2 \qquad (6.20)$$

$$\left(1-z^2x\right)\left(1-y^2x\right) \leq \left(1-xyz\right)^2 \qquad (6.21)$$

the multiplication of the left- and right-hand side of which yields an inequality closely related to the original inequality (6.6).

The simpler inequalities (6.19)–(6.21) are proved by direct manipulation. For example, proving inequality (6.19) is equivalent to proving the inequality $z^2y + x^2y \geq 2xyz$, which is equivalent to proving the inequality $z^2y + x^2y - xyz - xyz \geq 0$ or $zy(z-x) - xy(z-x) \geq 0$. The last inequality is equivalent to $(z-x)^2y \geq 0$ which is always true.

Because x, y and z are positive numbers between 0 and 1, the left- and right-hand sides of inequalities (6.19)–(6.21) are all non-negative. Consequently, multiplying the left and right-hand sides of inequalities (6.19)–(6.21) results in a correct inequality with the same direction:

$$\left[\left(1-x^2y\right)\left(1-y^2z\right)\left(1-z^2x\right)\right] \times \left[\left(1-y^2x\right)\left(1-z^2y\right)\left(1-x^2z\right)\right]$$
$$\leq \left[\left(1-xyz\right)\left(1-xyz\right)\left(1-xyz\right)\right]^2 \qquad (6.22)$$

However, inequality (6.22) can only be true if and only if both inequalities (6.23) and (6.24) are simultaneously true.

$$\left(1-x^2y\right)\left(1-y^2z\right)\left(1-z^2x\right)\le\left(1-xyz\right)\left(1-xyz\right)\left(1-xyz\right) \qquad (6.23)$$

$$\left(1-y^2x\right)\left(1-z^2y\right)\left(1-x^2z\right)\le\left(1-xyz\right)\left(1-xyz\right)\left(1-xyz\right) \qquad (6.24)$$

Again, note that the direction of both inequalities (6.23) and (6.24) cannot be '$\ge$' because, in such a scenario, the product of the left- and right-hand sides of inequalities (6.23) and (6.24) would result in an inequality with a direction opposite to that of inequality (6.22).

Suppose that only the direction of inequality (6.24) is reversed. As a result, according to our assumption, the inequality

$$\left(1-y^2x\right)\left(1-z^2y\right)\left(1-x^2z\right)\ge\left(1-xyz\right)\left(1-xyz\right)\left(1-xyz\right) \qquad (6.25)$$

holds. We aim to demonstrate that this particular assumption leads to a contradiction. To do so, we give the specific values $x=a$, $y=b$, $z=c$ and from inequality (6.23) we get:

$$\left(1-a^2b\right)\left(1-b^2c\right)\left(1-c^2a\right)\le\left(1-abc\right)\left(1-abc\right)\left(1-abc\right) \qquad (6.26)$$

Now let us interchange the values of variables x and y: $x=b$, $y=a$, $z=c$. From inequality (6.25) we get

$$\left(1-a^2b\right)\left(1-c^2a\right)\left(1-b^2c\right)\ge\left(1-abc\right)\left(1-abc\right)\left(1-abc\right) \qquad (6.27)$$

The left- and right-hand sides of inequalities (6.26) and (6.27) are identical; however, the direction of inequality (6.27) is opposite to that of inequality (6.26). This contradiction is a result of the assumption made about the direction of inequality (6.24). Consequently, the direction of inequality (6.24) is the same as the direction of inequality (6.23).

To avoid repetition, we will not restate the identical arguments used to prove the remaining inequalities (6.7) through (6.9). Moreover, the inequalities can readily be extended to encompass more than three variables.

It is important to highlight that the system reliability ranking of the series-parallel systems depicted in Figures 6.9–6.11 has been done regardless of the failure probabilities of the individual components building these systems or their ranking.

7 Reverse Engineering of Algebraic Inequalities to Disprove System Reliability Predictions Based on Average Component Reliablities

7.1 SYSTEM RELIABILITY ON DEMAND PREDICTIONS BASED ON AVERAGE COMPONENT RELIABILITIES

7.1.1 A GENERAL INEQUALITY RELATED TO SERIES-PARALLEL SYSTEMS

In this chapter, we use reverse engineering of a key algebraic inequality to demonstrate *that the prevalent practice of using average reliability on demand to calculate system reliability on demand, for components of the same type but different varieties, is fundamentally flawed.*

Consider the next valid algebraic inequality:

$$\left(1-x_1^m\right)\left(1-x_2^m\right)...\left(1-x_n^m\right) \le \left(1-\left[\left(x_1+x_2+...+x_n\right)/n\right]^m\right)^n \quad (7.1)$$

where $m \ge 1$ is an integer exponent and $x_1, ..., x_n$ are n real values for which $0 \le x_i \le 1$. A proof of this inequality has been provided in the Appendix.

7.1.2 SERIES-PARALLEL SYSTEMS

For $m = 2$, the general inequality (7.1) becomes

$$\left(1-x_1^2\right)\left(1-x_2^2\right)...\left(1-x_n^2\right) \le \left(1-\left[\left(x_1+x_2+...+x_n\right)/n\right]^2\right)^n \quad (7.2)$$

Inequality (7.2) can be reverse-engineered easily. Let x_i ($0 < x_i < 1$) be physically interpreted as the probability of failure of a component C_i of variety i,

DOI: 10.1201/9781003517764-7

where $i = 1, 2, ..., n$. Although all components are of the same type, they are inhomogeneous due to variations in properties, age and operating conditions.

The variables x_i are interpreted as *probability of failure on demand for components of variety i (all components are of the same type)*. Note that the time is not present in the probability of failure on demand. Also, the individual probabilities of failure on demand for components of different varieties but of the same type *are not known* in advance. Therefore, the use of average values for the probability of failure on demand in reliability predictions, *is inevitable*.

Because of differences in age, the number and size of material and manufacturing flaws, and differences in working conditions, no two components of the same type are identical in terms of reliability. There is no way of knowing the reliability of a component of type X from a particular variety (with a specified age, number, nature, size of the material flaws and manufacturing flaws, working conditions, etc.).

For example, the reliability variation of components is significantly influenced by the presence of material flaws, as well as their size, number density and location (Todinov, 2002a, 2006a). As a result, components of the same type and material, sourced from different suppliers, may exhibit considerable differences in their reliabilities. Since it is impossible to obtain the reliability on demand for components from different varieties, this intrinsic variability requires the use of average component reliability on demand (or probability of failure on demand). The average reliability on demand is what is listed in databases related to reliability on demand of components from a particular type. If, for example, 261 out of 900 valves of type X failed to close on command, the probability of failure on demand for valves of type X will be assessed by the average value of $261/900 = 0.29$.

An example of a system whose reliability depends on the probability of failure on demand of its components is given in Figure 7.1. The system consists of n pipelines transporting toxic fluid with two valves on each pipeline. All valves are initially open and a signal to close is sent to all valves in order to stop the fluid in each pipeline.

The system is deemed operational when, upon receiving a command for closure, the flow is halted in all n pipelines. To boost the system reliability, each pipeline features a redundant valve. This redundancy implies that at least one valve on each pipeline must respond to the closure command to ensure that the flow in the pipeline is halted.

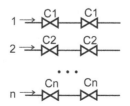

FIGURE 7.1 System of n pipelines transporting toxic fluid with two valves on each pipeline.

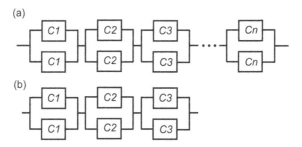

FIGURE 7.2 (a) Reliability network of a series-parallel system with components from n varieties; (b) Reliability network of a series-parallel system involving components of 3 varieties.

The valves are of varieties $C1$, $C2$,...,Cn, characterised by probabilities of failure to close on demand x_1, x_2, ..., x_n, correspondingly.

Consider the reliability network of the system in Figure 7.1 given in Figure 7.2a. It is a series-parallel system which is quite common in numerous engineering applications. For example, a system of n pipelines carrying toxic fluid, each equipped with a flange sealed by two seals one of which is redundant, could be an alternative to the system illustrated in Figure 7.1. Alternatively, instead of pipelines with flanges, we could consider a system consisting of n zones. In each zone, two sensors (one of which is redundant) measure the temperature, pressure, or concentration levels. A valid measurement from at least one sensor from each zone is necessary for the system to be operational.

The number of varieties of components of type X will be denoted by n. The left-hand part of inequality (7.2) is the actual reliability of the system in Figure 7.2a. The expression $\bar{x} = \dfrac{x_1 + x_2 + ... + x_n}{n}$ in the right-hand part of (7.2) is the average probability of failure $\bar{x}$ on demand for the varieties of the selected type X (e.g. valve), assessed as an average related to n varieties.

Note that the probabilities of failure x_i characterising the n varieties are not known and this is why the system reliability on demand cannot be estimated directly, by using these probabilities. Because the expression for $\bar{x}$ cannot be evaluated using x_i, the ratio p_f/p is always used instead where p_f is the number of observed in the past failed components from the type X and p is the total number of observed components from that type. Note that p_f and p are obtained from component failure statistics and *are not related* to the number of components building the system.

The numbers of p_f and p should be sufficiently large to produce an accurate estimate of the probability of failure on demand p_f/p. Thus, for a system with n components of the same type X and of different variety (n varieties in total), for the average probability of failure $\bar{x}$ of the components from that particular type X, the following equation holds:

$$\bar{x} = \frac{x_1 + x_2 + ... + x_n}{n} \approx \frac{p_f}{p} \qquad (7.3)$$

Equation (7.3) can be proved by considering that the left-hand side of (7.3) can be presented as

$$\frac{x_1 + x_2 + \ldots + x_n}{n} = \left(1/n\right)x_1 + \left(1/n\right)x_2 + \ldots + \left(1/n\right)x_n \tag{7.4}$$

This essentially represents the total probability associated with the failure of a component in the system. Indeed, a component from type X, can fail in n mutually exclusive ways. This includes the scenario where the component belongs to variety 1 and fails (a compound event with probability $(1/n)x_1$), the scenario where the component belongs to variety 2 and fails (a compound event with probability $(1/n)x_2$), and so on. The probability of failure of a component in the system must approach p_f/p because this ratio is the empirical probability of failure for a component of type X.

Very similar reasoning also applies to the case where the number n of component varieties is smaller than the number n_c of components in the system ($n < n_c$).

Indeed, let $n_1, n_2, \ldots, n_n$ $\left(\sum_{i=1}^{n} n_i = n_c \right)$ be the number of components in the system from each variety (these numbers are also unknown). The total probability of failure for a component in the system is then given by

$$\frac{p_f}{p} \approx \frac{n_1 x_1 + n_2 x_2 + \ldots + n_n x_n}{n_c} \tag{7.5}$$

where p_f is the observed in the past total number of failed components (from failure statistics) and p is the total number of observed components in the past. The right-hand side of (7.5) is the weighted average of the probabilities of failure characterising the n varieties.

Indeed, a component in the system can fail in n mutually exclusive ways. This includes the scenario where the component belongs to variety 1 and fails (a compound event with probability $(n_1/n_c)x_1$, the scenario where the component belongs to variety 2 and fails (a compound event with probability $(n_2/n_c)x_2$), and so on.

The total probability of a component failure is then given by the expression:

$$\left(n_1/n_c\right)x_1 + \ldots + \left(n_n/n_c\right)x_n$$

which must approach the empirical probability p_f/p of component failure and the result is (7.5).

To test (7.3) and (7.5), Monte Carlo simulations were performed, based on $p = 100000$ observed components and $n = 1, 2, \ldots, 10$, component varieties. In an array, random values between 0 and 1 are initially assigned for the probabilities of failure characterising the n varieties. Next, $p = 100000$ components were selected by choosing randomly their variety. Each randomly selected component was also virtually tested for failure on demand by using the probability of failure on demand characterising its variety. At the end of the simulation, the ratio of the total number of failed components p_f and the total number $p = 100000$ observed components was formed. The validity of (7.3) and (7.5) has been confirmed with each Monte Carlo simulation.

Inequality (7.2) can also be rewritten as

$$\left(1-\bar{x}^2\right)^n \geq \left(1-x_1^2\right)\left(1-x_2^2\right)\ldots\left(1-x_n^2\right) \tag{7.6}$$

or considering (7.5), it can also be rewritten as:

$$\left(1-\left(p_f/p\right)^2\right)^n \geq \left(1-x_1^2\right)\left(1-x_2^2\right)\ldots\left(1-x_n^2\right) \tag{7.7}$$

The right-hand side of inequality (7.7) gives the actual reliability on demand of the series-parallel system in Figure 7.2a including components of n varieties and of the same type. Because the probabilities of failure x_i characterising the separate varieties and the number of components from the separate varieties are not known, the system reliability on demand must necessarily be estimated through the left-hand side of inequality (7.7).

As a result, the reverse engineering of inequality (7.2) yields that the predicted reliability on demand of a series-parallel systems based on an average probability of failure on demand $\bar{x} = p_f / p$, is higher than the actual reliability of the system. This always holds true provided that the estimate $\bar{x} = p_f / p$ is sufficiently accurate (Todinov, 2024b). The significant divergence between the projected and actual system reliability on demand, caused by variability, can be remarkably pronounced, as evidenced by the next numerical examples.

Let's consider 900 valves of the same type X but of three different varieties (for example, valves from machine centres 1, 2 and 3). The valves work independently from one another. From past failure statistics, 261 of the monitored 900 valves fail to close on demand. Because only the total number of valves 900 and the total number of unreliable valves are known, the probability of failure on demand for the valves of type X will be estimated from:

$$\bar{x} = p_f / p = 261 / 900 = 0.29$$

Now, suppose that the series-parallel system in Figure 7.2b includes two valves from each of the three varieties. The estimated system reliability based on average probability of failure on demand $\bar{x}$ becomes:

$$R_{\text{est}} = \left(1-\left(p_f/p\right)^2\right)^3 = \left(1-0.29^2\right)^3 = 0.77.$$

For the sake of simplicity, assume that each of the three manufacturing centres has produced 300 valves of type X, resulting in valves of three distinct varieties. The total number of valves produced by the manufacturing centres is 900. Let the number of unreliable valves from these varieties (manufacturing centres) be 12, 42, and 207, respectively (the total number of manufactured unreliable valves is 261). Consequently, the probability of failure on demand for each variety is as follows:

$$x_1 = 12/300 = 0.04, \ x_2 = 42/300 = 0.14 \text{ and } x_3 = 207/300 = 0.69, \text{ correspondingly.}$$

As can be verified, the following expression holds true for the average probability of failure $\bar{x}$:

$$\bar{x} = (x_1 + x_2 + x_3)/3 = (0.04 + 0.14 + 0.69)/3 = 0.29 = p_f / p = 261/900$$

Suppose that valves from each variety have been used to construct the three sections arranged in series in Figure 7.2b. The actual (real) reliability of the series-parallel system is:

$$R_{\text{real}} = (1 - 0.04^2) \times (1 - 0.14^2) \times (1 - 0.69^2) = 0.51$$

The estimated reliability on demand $R_{\text{est}} = 0.77$ is 1.51 times greater than the real reliability $R_{\text{real}} = 0.51$!

In the next example, the actual reliability of the system in Figure 7.3a is given by the left-hand part of inequality (7.1) for m components in each section in series, while the right-hand part provides an estimate of the system reliability based on

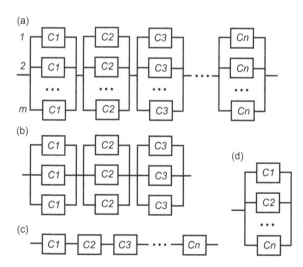

FIGURE 7.3 (a) Reliability network of a series-parallel system with components from n varieties and m components in each block; (b) Reliability network of a series-parallel system with components from 3 varieties and 3 components in each block; (c) Reliability network of a system with components in series, from the same type and n varieties; (d) Reliability network of a system with components in parallel, from the same type and n varieties.

the average probability of failure on demand characterising the n different component varieties.

For $n = 3$ sections with $m = 3$ redundant components in each section (Figure 7.3b), inequality (7.1) becomes

$$\left(1-x_1^3\right)\left(1-x_2^3\right)\left(1-x_3^3\right) \le \left(1-\left[\left(x_1+x_2+x_3\right)/3\right]^3\right)^3 = \left(1-\left(p_f/p\right)^3\right)^3 \quad (7.8)$$

where the left-hand part of (7.8) is the actual reliability of the system in Figure 7.3b while the right-hand part is the reliability of the system in Figure 7.3b, estimated by using the average probability of failure on demand of the three varieties.

For components of the same three varieties as in the previous example, the left-hand side of inequality (7.8) gives:

$$R_{\text{real}} = \left(1-0.04^3\right)\times\left(1-0.14^3\right)\times\left(1-0.69^3\right) = 0.67$$

for the real reliability of the arrangement in Figure 7.3b.

If the reliability of the section in Figure 7.3b is calculated on the basis of the average probability of failure on demand $\bar{x} = p_f / p = 261/900 = 0.29$ characterising the three varieties, for the estimated system reliability on demand, the right-hand side of (7.8) gives a significantly larger value:

$$R_{\text{est}} = \left(1-\left(p_f/p\right)^3\right)^3 = \left(1-0.29^3\right)^3 = 0.93.$$

7.1.3 System with Components Logically Arranged in Series

If in inequality (7.1), we set $m = 1$, the inequality transforms into:

$$\left(1-\bar{x}\right)^n \ge \left(1-x_1\right)\left(1-x_2\right)\ldots\left(1-x_n\right) \quad (7.9)$$

where $\bar{x} = p_f / p$ is the average probability of failure on demand. For the sake of simplicity, for systems in series, let us assume that the number of components is equal to the number n of varieties. Clearly, the right-hand part of inequality (7.9) is the actual reliability on demand of a system with n components logically arranged in series, of the same type X and n different varieties [see Figure 7.3c]. The left-hand part of inequality (7.9) is the reliability on demand of the same system, evaluated by taking an average probability of failure on demand $\bar{x} = p_f / p$ for the components. The quantity $r_i = 1 - x_i$ in inequality (7.9) is the reliability on demand of the components from variety i ($i = 1,\ldots,n$) while $\bar{r} = 1 - \bar{x}$ is the average reliability on demand characterising all components of the given type X. Noticing that

$$1 - \bar{x} = 1 - \frac{x_1 + x_2 + \ldots + x_n}{n} = \frac{(1 - x_1) + (1 - x_2) + \ldots + (1 - x_n)}{n}$$

inequality (7.9) can also be rewritten as

$$\left(1 - \frac{p_f}{p}\right)^n = \left(\frac{r_1 + r_2 + \ldots + r_n}{n}\right)^n \geq r_1 r_2 \ldots r_n \qquad (7.10)$$

which is the classical Arithmetic mean – Geometric mean (AM-GM) inequality (Steele, 2004).

Inequality (7.10) clearly highlights the overestimation of system reliability on demand when using the average reliability on demand for components of different varieties. This conclusion could have been reached directly, without addressing the special case of inequality (7.9), if reverse engineering had been applied to the AM-GM inequality (7.10).

Consider the same three varieties as in the previous example, with the valves logically arranged in series. As in the previous examples, the varieties have reliabilities on demand of $0.96 = 1 - 0.04$, $0.86 = 1 - 0.14$ and $0.31 = 1 - 0.69$, respectively, and the valves operate independently of one another.

The right-part of inequality (7.10) gives the actual reliability of the series arrangement, which is a product of the reliabilities on demand of the independently operating valves. Therefore, the actual reliability of a system based on three valves is:

$$R_{\text{real}} = 0.96 \times 0.86 \times 0.31 = 0.25$$

The left-hand part of inequality (7.10) gives the system reliability calculated on the basis of the average reliability on demand $\bar{r}$ characterising the three varieties of valves:

$$\bar{r} = 1 - p_f / p = 1 - 261 / 900 = 0.71$$

Note that $\bar{r} = (0.96 + 0.86 + 0.31) / 3 = 0.71$.

The estimated system reliability based on average reliability on demand from the left part of inequality (7.10) is:

$$R_{\text{est}} = \bar{r}^3 = 0.71^3 = 0.36.$$

The estimated value $R_{\text{est}} = 0.36$ is 1.44 times greater than the real reliability $R_{\text{real}} = 0.25$ of the section!

7.1.4 System with Components Logically Arranged in Parallel

For the sake of simplicity, for systems in parallel, we also assume that the number of components is equal to the number n of varieties. Suppose that x_i in the AM-GM inequality are interpreted to be the probabilities of failure x_i of the different component varieties. From the AM-GM inequality we then have:

$$\left(p_f / p\right)^n = \left(\frac{x_1 + x_2 + \ldots + x_n}{n}\right)^n \geq x_1 x_2 \ldots x_n \qquad (7.11)$$

The right-hand side of inequality (7.11) then can be interpreted as the probability of failure of a system where all independently working components are logically arranged in parallel (Figure 7.3d). A system consisting of n independently working components arranged in parallel is in a failed state when all components are in a failed state. The left-hand part of inequality (7.11) is the probability of failure on demand of the system in parallel, estimated on the basis of the average probability of failure on demand of the components from different varieties. The physical interpretation of inequality (7.11) suggests that for parallel systems, when estimates are derived from the average probability of component failure on demand, the projected probability of system failure exceeds the actual value.

In summary, even for statistically independent components, the conventional methods considered in this chapter are not reliable methods for assessing the system reliability on demand. Based on the findings from the reverse engineering of inequality (7.1), we can draw the following conclusions. *The prevailing methodology for predicting system reliability on demand, which relies on average component reliabilities on demand for components of different varieties but the same type, is fundamentally flawed due to component variability* (Todinov, 2024b).

If there were no variability in the reliabilities of components of the same type, inequalities (7.10–7.11) would become equalities, and there would be no discrepancy between the estimated and the actual system reliability. The larger the deviations of the component reliabilities from the average value, the stronger the inequalities (7.10–7.11) will be. For any component, deviations of its reliability on demand from the average value are inevitable, primarily due to differences in age, manufacturers, working conditions, presence of material, and manufacturing flaws. Consequently, discrepancies between the predicted reliability on demand and the actual value will always exist.

Assuming average reliability on demand for components of a particular type, however, can still provide valuable insights, if the research scope is confined to a comparative analysis that ranks competing system designs.

Thus, the new knowledge generated from the reverse engineering of inequalities (7.1) and (7.2) helped to expose a fundamental flaw in the existing practice of predicting reliability based on the average reliability values characterizing components of different types.

7.2   EVALUATING SYSTEM RELIABILITY RELATED TO A SPECIFIED TIME INTERVAL

It is important to discuss also the impact of variability on reliability predictions when evaluating reliability over a specified time interval.

Consider a system with components of the same type and n varieties, logically arranged in parallel. Suppose that each component variety is characterised by a constant hazard rate λ_i, $i = 1, \ldots, n$. Consequently, the time to failure distribution for a component is the negative exponential time to failure distribution (Ramakumar 1993; Ebeling, 1997; Hoyland and Rausand 1994):

$$F(t) = 1 - \exp(-\lambda_i t) \tag{7.12}$$

where t is the time interval and $F(t)$ is the probability that the component will fail before time t.

Consider n components logically arranged in parallel (Figure 7.3d), each of which is from the same type but from a different variety. According to the system reliability theory (Ramakumar 1993; Ebeling, 1997; Hoyland and Rausand 1994), the actual probability of failure F_{s1} of the system before time t is given by

$$F_{s1} = \left(1 - e^{-\lambda_1 t}\right) \times \left(1 - e^{-\lambda_2 t}\right) \times \ldots \times \left(1 - e^{-\lambda_n t}\right) \tag{7.13}$$

(The system is in a failed state at time t only if all components are in a failed state at time t).

If the probability of failure or the system reliability is calculated based on the average hazard rate:

$$\bar{\lambda} = \frac{\lambda_1 + \lambda_2 + \ldots + \lambda_n}{n} \tag{7.14}$$

the probability of system failure before time t is given by

$$F_{s2} = \left(1 - e^{-\bar{\lambda} t}\right)^n \tag{7.15}$$

It can be shown that the inequality:

$$\left(1 - e^{-\bar{\lambda} t}\right)^n \geq \left(1 - e^{-\lambda_1 t}\right) \times \left(1 - e^{-\lambda_2 t}\right) \times \ldots \times \left(1 - e^{-\lambda_n t}\right) \tag{7.16}$$

always holds. Inequality (7.16) can be proved by taking logarithms from both sides, showing that the right-hand side is a concave function and applying the Jensen's inequality. The proof of inequality (7.16) is very similar to the proof of inequality (7.1) given in the Appendix and details will be omitted. As a result, using average hazard rate to estimate the reliability of systems in parallel always leads to overestimating the probability of failure of the system (Todinov, 2024b).

Despite their popularity, system reliability predictions based on average failure rates (e.g., MIL-STD-1629A, 1977) have very serious shortcomings. The failure of this approach to generate accurate system reliability predictions led to growing disillusionment among researchers and practitioners. Not surprisingly, many researchers and practitioners abandoned the use of failure rate-based reliability predictions.

For systems with components of the same type, logically arranged in series, the system reliability estimated on the basis of an averaged hazard rate is exactly equal to the actual reliability of the system.

Indeed, let $\lambda_1, \lambda_2, \ldots, \lambda_n$ be the constant hazard rates characterising the n component varieties. Since the reliability of a single component of variety i is given by $R_i = \exp(-\lambda_i t)$, the reliability of the system based on n components of the same type, each of which is of different variety, is given by the expression

$$R_{s1} = \exp\left(-\lambda_1 t\right) \times \exp\left(-\lambda_2 t\right) \times \ldots \times \exp\left(-\lambda_n t\right) = \exp\left(-\left[\lambda_1 + \lambda_2 + \ldots + \lambda_n\right] t\right) \quad (7.17)$$

If the reliability of the system in series is calculated on the basis of the average hazard rate of the n varieties (given by (7.14)), the system reliability is given by

$$R_{s2} = \exp\left(-\bar{\lambda} t\right) \times \exp\left(-\bar{\lambda} t\right) \times \ldots \times \exp\left(-\bar{\lambda} t\right) = \exp\left(-\left[\lambda_1 + \lambda_2 + \ldots + \lambda_n\right] t\right) \quad (7.18)$$

Since $R_{s1} = R_{s2}$, in this case, working with an average hazard rate for components of a particular type neither overestimates nor underestimates the predicted system reliability.

APPENDIX 7.A: PROOF OF INEQUALITY (7.1)

From the basic properties of the concave functions $f(x)$ and $g(x)$: $f[\lambda x + (1 - \lambda)y] \geq \lambda f(x) + (1 - \lambda)f(y)$, and $g[\lambda x + (1 - \lambda)y] \geq \lambda g(x) + (1 - \lambda)g(y)$ where $0 \leq \lambda \leq 1$, it can be shown easily that the sum $h(x) = f(x) + g(x)$ of two concave functions $f(x)$ and $g(x)$ is a concave function. By induction, it can be deduced that the sum of n concave functions is also a concave function.

Inequality (7.1) can be proved by observing that the sum of the logarithms: $z = \ln\left(1 - x_1^m\right) + \ln\left(1 - x_2^m\right) + \ldots + \ln\left(1 - x_n^m\right)$ is a concave function because it is a sum of the n concave functions $z_1 = \ln\left(1 - x_1^m\right); z_2 = \ln\left(1 - x_2^m\right); \ldots; z_n = \ln\left(1 - x_n^m\right)$. The functions $z_i = \ln\left(1 - x_i^m\right)$ are concave because their second derivatives are all negative:

$$\frac{\partial^2 z_i}{\partial x_i^2} = -\frac{m(m-1)x_i^{m-2}\left(1-x_i^m\right)+m^2 x_i^{2(m-1)}}{\left(1-x_i^m\right)^2} < 0.$$

considering that $m - 1 \geq 0$ and $1 - x_i^m > 0$.

Let w_i be weights defined such that $w_1 = w_2 = \ldots = w_n = 1/n$. According to the Jensen's inequality (Steele, 2004), if $\bar{x} = w_1 x_1 + w_2 x_2 + \ldots + w_n x_n$, the following inequality holds for a concave function:

$$w_1 \times \ln\left(1-x_1^m\right) + w_2 \times \ln\left(1-x_2^m\right) + \ldots + w_n \times \ln\left(1-x_n^m\right)$$
$$\leq \ln\left(1-\left(w_1 x_1 + w_2 x_2 + \ldots + w_n x_n\right)^m\right) \tag{7.A.1}$$

Inequality (7.A.1) can then be rewritten as:

$$\ln\left[\left(1-x_1^m\right)\left(1-x_2^m\right)\ldots\left(1-x_n^m\right)\right] \leq \ln\left[1-\left(\left(x_1 + x_2 + \ldots + x_n\right)/n\right)^m\right]^n \tag{7.A.2}$$

Since the exponential function e^x is strictly increasing, according to the properties of inequalities, the direction of inequality (7.A.2) will not change if both sides of (7.A.2) are exponentiated:

$$\exp\left(\ln\left[\left(1-x_1^m\right)\left(1-x_2^m\right)\ldots\left(1-x_n^m\right)\right]\right)$$
$$\leq \exp\left(\ln\left[1-\left(\left(x_1 + x_2 + \ldots + x_n\right)/n\right)^m\right]^n\right) \tag{7.A.3}$$

which yields inequality (7.1). This completes the proof.

8 Reverse Engineering of the Inequality of Additive Ratios

8.1 INEQUALITY OF THE ADDITIVE RATIOS

Consider the algebraic inequality

$$\frac{p_1}{q_1} + \frac{p_2}{q_2} + \ldots + \frac{p_n}{q_n} \geq n \frac{p_1 + p_2 + \ldots + p_n}{q_1 + q_2 + \ldots + q_n} \tag{8.1}$$

where $p_1 \geq p_2 \geq \ldots \geq p_n$ and $q_1 \leq q_2 \leq \ldots \leq q_n$ are positive additive quantities.

Inequality (8.1) will be referred to as *the inequality of the additive ratios* (Todinov, 2022b). Equality in (8.1) is attained for proportional p_i and q_i ($p_i/q_i = p_j/q_j = k$). Indeed, the substitution $p_i = kq_i$ in inequality (8.1) yields a value nk for both the left-hand side and the right-hand side. If there are i and j ($i \neq j$) for which $p_i/q_i \neq p_j/q_j$, the left-hand side of (8.1) is strictly greater than the right-hand side.

Inequality (8.1) can be proved by using the classic Chebyshev's sum inequality (see Section 2.1.6) and the arithmetic mean-harmonic mean inequality (see Section 2.1.4).

According to the classic Chebyshev's sum inequality, for two similarly ordered sequences of positive numbers: $p_1 \geq p_2 \geq \ldots \geq p_n$ and $b_1 \geq b_2 \geq \ldots \geq b_n$, the following inequality holds:

$$\frac{p_1 b_1 + p_2 b_2 + \ldots + p_n b_n}{n} \geq \frac{p_1 + p_2 + \ldots + p_n}{n} \times \frac{b_1 + b_2 + \ldots + b_n}{n} \tag{8.2}$$

Let us make the setting $b_i = 1/q_i$, $i = 1,\ldots,n$. For the positive sequences $p_1 \geq p_2 \geq \ldots \geq p_n$ and $1/q_1 \geq 1/q_2 \geq \ldots \geq 1/q_n$, applying the Chebyshev's inequality (8.2) gives

$$\frac{p_1/q_1 + p_2/q_2 + \ldots + p_n/q_n}{n} \geq \frac{p_1 + p_2 + \ldots + p_n}{n} \times \frac{1/q_1 + 1/q_2 + \ldots + 1/q_n}{n} \tag{8.3}$$

DOI: 10.1201/9781003517764-8

According to the arithmetic mean – harmonic mean inequality, for any sequence $y_1, y_2, \ldots, y_n$ of real numbers, the inequality

$$\frac{y_1 + y_2 + \ldots + y_n}{n} \geq \frac{n}{1/y_1 + 1/y_2 + \ldots + 1/y_n} \tag{8.4}$$

holds. From inequality (8.4), the inequality

$$\frac{1/y_1 + 1/y_2 + \ldots + 1/y_n}{n} \geq \frac{n}{y_1 + y_2 + \ldots + y_n} \tag{8.5}$$

can be obtained. If inequality (8.5) is applied to the expression $\dfrac{1/q_1 + 1/q_2 + \ldots + 1/q_n}{n}$ in inequality (8.3), the result is the inequality:

$$\frac{1/q_1 + 1/q_2 + \ldots + 1/q_n}{n} \geq \frac{n}{q_1 + q_2 + \ldots + q_n}$$

Substituting the expression $\dfrac{1/q_1 + 1/q_2 + \ldots + 1/q_n}{n}$ in the right-hand side of inequality (8.3) with the smaller quantity $\dfrac{n}{q_1 + q_2 + \ldots + q_n}$, only strengthens inequality (8.3). After substituting and cancelling n, inequality (8.1) is obtained.

By using inequality (8.1), the effect of the additive quantities $p = p_1 + \ldots + p_n$ and $q = q_1 + \ldots + q_n$ can be increased more than n times by dividing them into smaller segments p_i, and q_i ($i = 1, \ldots, n$). The application of inequality (8.1) however, requires the ratios p_i/q_i to be additive quantities, which, in some cases is not fulfilled. Indeed, if p stands for a particular additive quantity and q stands for another additive quantity, the ratio p/q can often be non-additive. Nevertheless, in some of these cases, inequality (8.1) can still be physically interpreted if it is multiplied by an appropriate positive constant λ. The result is that the non-additive quantities p_i/q_i now become additive quantities $\lambda p_i/q_i$. In this case, inequality (8.1) becomes

$$\frac{\lambda p_1}{q_1} + \frac{\lambda p_2}{q_2} + \ldots + \frac{\lambda p_n}{q_n} \geq n\lambda \frac{p_1 + p_2 + \ldots + p_n}{q_1 + q_2 + \ldots + q_n} \tag{8.6}$$

Inequality (8.6) is *the inequality of the additive ratios*: *if p and q are additive quantities, and if the output quantity λp/q is also an additive quantity (where λ is a constant), then, segmenting the quantities p and q into n asymmetric parts increases by more than n times the total output quantity* (Todinov, 2022b).

8.2 REVERSE ENGINEERING OF THE INEQUALITY OF THE ADDITIVE RATIOS

8.2.1 IMPROVING THE PROBABILITY OF AN EVENT OCCURRING WITH A NUMBER OF MUTUALLY EXCLUSIVE EVENTS

The inequality of the additive ratios (8.1) can be reverse engineered in the probability domain and has powerful probabilistic applications. To reverse engineer inequality (8.1) in terms of probabilities, it is presented as:

$$\left(1/n\right)\frac{p_1}{q_1} + \left(1/n\right)\frac{p_2}{q_2} + ... + \left(1/n\right)\frac{p_n}{q_n} \geq \frac{p_1 + p_2 + ... + p_n}{q_1 + q_2 + ... + q_n} \tag{8.7}$$

where $p_1 \geq p_2 \geq ... \geq p_n$ and $q_1 \leq q_2 \leq ... \leq q_n$.

Consider an event B occurring with a set of n mutually exclusive events A_i (Figure 8.1). Each event A_i is selected (occurs) with the same probability $1/n$. Given that event A_i has occurred, let p_i in inequality (8.1) be physically interpreted as the number of favourable outcomes leading to event B. Let q_i be physically interpreted as the total number of outcomes for event A_i. Because $(1/n)$ is the probability of selecting event A_i, the ratio $\left(1/n\right)\frac{p_i}{q_i}$ is the probability of event B given that event A_i has been selected.

The left-hand side of inequality (8.7) now gives the probability of event B, which occurs after the random selection of an event A_i (with a uniform probability $1/n$).

The right-hand side of inequality (8.7) gives the probability of event B, occurring as a result of $\sum_i p_i$ total number of favourable outcomes out of $\sum_i q_i$ total number of possible outcomes.

At first sight, it seems that inequality (8.7) is a mere restatement of the well-known total probability theorem and a strict inequality cannot possibly hold. This first impression however, is incorrect. If event B occurs with n mutually exclusive and exhaustive events A_i, the well-known total probability theorem (DeGroot, 1989) states that

$$P\left(B\right) = P\left(A_1\right) \times P(B \mid A_1) + P\left(A_2\right) \times P(B \mid A_2) + ... + P\left(A_n\right) \times P(B \mid A_n) \tag{8.8}$$

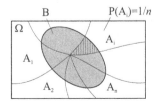

FIGURE 8.1 Mutually exclusive events A_i, each selected with probability $1/n$ and event B, occurring with one of the events A_i.

Substituting $P(A_i) = 1/n$ and $P(B|A_i) = p_i/q_i$ in Equation (8.8) indeed gives the left-hand side

$$(1/n)\frac{p_1}{q_1}+(1/n)\frac{p_2}{q_2}+...+(1/n)\frac{p_n}{q_n}$$

of inequality (8.7).

However, the left-hand side of (8.7) *is not*, in general, equal to the right-hand side. If event A_i in equation (8.8) occurred with probabilities $P(A_i)$ proportional to the number of outcomes q_i leading to the event A_i, the left-hand side of equation (8.7) would be equal to the right-hand side $\frac{p_1+p_2+...+p_n}{q_1+q_2+...+q_n}$. Indeed, if $P(A_i)=\dfrac{q_i}{\sum_i q_i}$ then, according to the total probability theorem,

$$P(B)=\frac{q_1}{\sum_i q_i}\times\frac{p_1}{q_1}+\frac{q_2}{\sum_i q_i}\times\frac{p_2}{q_2}+...+\frac{q_n}{\sum_i q_i}\times\frac{p_n}{q_n}=\frac{p_1+p_2+...+p_n}{q_1+q_2+...+q_n}$$

In interpreting inequality (8.7), however, the events A_i are selected *with the same probability* $1/n$, irrespective of the number of outcomes q_i associated with the events A_i. If events A_i are deliberately selected with the same probability $P(A_i) = 1/n$, and if there are i and j ($i \ne j$) for which $p_i/q_i \ne p_j/q_j$, then a strict inequality holds:

$$P(B)=(1/n)\frac{p_1}{q_1}+(1/n)\frac{p_2}{q_2}+...+(1/n)\frac{p_n}{q_n} > \frac{p_1+p_2+...+p_n}{q_1+q_2+...+q_n} \qquad (8.9)$$

The strict inequality (8.9) means that there is indeed an increase in the probability of occurrence of event B.

Despite decades of research in probability theory, this key result remained elusive to probability experts which demonstrates the power of reverse engineering of algebraic inequalities.

8.2.2   INCREASING THE PROBABILITY OF PURCHASING A RELIABLE PRODUCT

A useful application of inequality (8.7) can be given with n suppliers of the same type of product. Suppose that each supplier delivers the same number p reliable products in the market. Alongside the reliable products, however, the suppliers also sell products with inferior reliability. The exact numbers of products with inferior reliability sold by the suppliers are not known. As a result, the total numbers of products delivered by the separate suppliers, given by q_i, $i = 1, ..., n$ are unknown. Let B denote the event 'purchasing a reliable product'. The question of interest is which strategy is more beneficial: purchasing a product from a randomly selected supplier or purchasing a product from the market formed by the n suppliers.

The first impression is that there should not be any difference. After all, the total numbers of products delivered by the individual suppliers on the market are unknown and the number of reliable products delivered by each supplier is the same – equal to p. This impression, however, is false as the next argument clearly demonstrates.

Suppose that a supplier is selected randomly, with the same probability $(1/n)$, followed by purchasing a product from that supplier. *According to inequality (8.7), this strategy is associated with a higher probability of purchasing a reliable product compared to purchasing a random product from the market formed by the n suppliers* (Todinov, 2023d).

The probability of purchasing a reliable product given that the ith supplier has been selected is given by the ratio p_i/q_i where $p_i = p$ is the number of reliable products from the ith supplier and q_i is the total number of products sold by the ith supplier.

The conditions for the validity of inequality (8.7) are fulfilled automatically because the supplied from the individual suppliers numbers of products q_i can always be ordered in ascending order $q_1 \leq q_2 \leq, \ldots, \leq q_n$ and the numbers p_i of reliable products is the same for each supplier, $p_1 = p_2 =, \ldots, = p_n = p$ and, as a result, $p_1 \geq p_2 \geq, \ldots, \geq p_n$ holds automatically.

The left-hand side of inequality (8.7) can then be interpreted as the total probability that a reliable product will be purchased if a supplier is selected with a uniform probability and a product is purchased from that supplier. Purchasing from suppliers are mutually exclusive events. Consequently, the probability that a reliable product will be purchased after randomly selecting a supplier, followed by purchasing a product from the selected supplier, is equal to the sum of the probabilities $(1/n)\dfrac{p_i}{q_i}$ of n mutually exclusive events.

The right-hand side of inequality (8.7) is the probability of purchasing a reliable product from the market formed by all n suppliers. Inequality (8.7) therefore, yields a highly counter-intuitive result. For suppliers delivering the same quantities of reliable products alongside unknown quantities of products of inferior reliability, the probability of purchasing a reliable product from a randomly selected supplier is higher than the probability of purchasing a reliable product from the market formed by all suppliers.

The seemingly plausible argument that this is because the suppliers with a small number of products q_i are characterised by a higher chance p_i/q_i of selling a reliable product and are selected more often, does not explain the result. This is because the diametrically opposite statement can also be made: the suppliers with a large number of products q_i are characterised by a small probability p_i/q_i of selling a reliable product and are still selected with the same probability $1/n$ as other suppliers. These suppliers have the same chance of being selected, followed by purchasing a product with inferior reliability. It is not clear how these two opposite effects interact and which effect is dominating.

The difference between the probabilities associated with these alternative strategies is significant. The deliberately simplified example given next, illustrates this significant difference. Suppose that three suppliers deliver $q_1 = 135$; $q_2 = 34$; $q_3 = 79$ numbers of products and each supplier delivers $p_1 = p_2 = p_3 = 20$ number of reliable products, correspondingly. Applying inequality (8.7) for $n = 3$ gives

$$(1/3)\frac{20}{135} + (1/3)\frac{20}{34} + (1/3)\frac{20}{79} = 0.33 > \frac{20+20+20}{135+34+79} = 0.24$$

which is an increase of the probability (0.24) of selecting a reliable component by 37.5%.

Inequality (8.7) has also been confirmed by a Monte-Carlo simulation.

Note that if the total number of products delivered by the suppliers was the same (instead of the number of reliable products being the same), no increase of the probability of purchasing a reliable product would be present.

In inequality (8.7), equality is attained if, for example, the numbers of products delivered by the suppliers are equal despite that the number of reliable products delivered by the separate suppliers may differ. Indeed, if $q_1 = q_2 = \ldots = q_n = q/n$, ($p_1 > p_2 > \ldots > p_n$), inequality (8.7) transforms into equality.

Irrespective of the differences in the 'density' of reliable products characterising the suppliers, the probability of purchasing a reliable product from a randomly selected supplier always remains equal to the probability of purchasing a reliable product in the market formed by these suppliers, as long as the number of products $q_1, q_2, \ldots, q_n$ delivered by the separate suppliers is the same.

8.2.3 INCREASING THE MASS OF SUBSTANCE DEPOSITED DURING ELECTROLYSIS AND AVOIDING OVERESTIMATION OF DENSITY

Consider the inequality of the additive ratios

$$\frac{a_1}{x_1} + \frac{a_2}{x_2} + \ldots + \frac{a_n}{x_n} \geq n\frac{a_1 + a_2 + \ldots + a_n}{x_1 + x_2 + \ldots + x_n} \tag{8.10}$$

where $a_1 \geq a_2 \geq \ldots \geq a_n$ and $x_1 \leq x_2 \leq \ldots \leq x_n$ are positive numbers.

In order to reverse-engineer inequality (8.10) in terms of a physical process, the ratios a_i/x_i of the additive quantities a_i and x_i must also be additive physical quantities.

If a_i/x_i in (8.10) is an additive quantity and a_i, x_i are in turn additive quantities, inequality (8.10) provides the mechanism to increase at least n times the effect of the aggregated additive quantities $a = \sum_{i=1}^{n} a_i$ and $x = \sum_{i=1}^{n} x_i$ by segmenting them into smaller parts a_i and x_i, $i = 1, \ldots, n$ and accumulating their individual effects a_i/x_i. Inequality (8.10) can always be reverse engineered as long as a_i, x_i and the terms a_i/x_i are additive quantities and have meaningful physical interpretation.

An alternative way of formulating this requirement is to be able to present the additive quantity a_i as a product of the additive quantity p_i and the additive quantity x_i:

$$a_i = p_i \times x_i \qquad (8.11)$$

It is important to note that inequality (8.10) can be applied to each of the individual terms a_i/x_i on the left-hand side which in turn can be segmented. The result from this recursive segmentation is a multiplication of the effects from the segmentation.

Inequality (8.10) will be reverse engineered to provide a basis for increasing the mass of substance deposited on electrodes during electrolysis. The Faraday's first law of electrolysis (Wolfson, 2016) states that the mass m of substance deposited at an electrode in grams is directly proportional to the charge Q in Coulombs:

$$m = Z \times Q \qquad (8.12)$$

where Z is a constant of proportionality called electro-chemical equivalent of the substance. It is the mass deposited for a charge of 1 Coulomb. By using the relationship between charge Q, current I [A] and time t in seconds, equation (8.12) can be rewritten as

$$m = Z \times I \times t \qquad (8.13)$$

which gives the mass m of substance deposited at an electrode in grams by a current of magnitude I, for a time of t seconds.

Consider now a process of electrolysis induced by voltage of magnitude V applied to a cell with resistance R. Since the current I is determined from the Ohm's law

$$I = V / R \qquad (8.14)$$

the mass m_0 of deposited substance is given by

$$m_0 = Z \times (V / R) \times t \qquad (8.15)$$

Consider an alternative design, for which the electrolysis is conducted after segmenting the initial cell with resistance R into two smaller cells with smaller resistances r_1 and r_2 ($r_1 + r_2 = R$; $r_1 \leq r_2$). The voltages applied to the cells are with magnitudes V_1 and V_2 ($V_1 \geq V_2$ and $V_1 + V_2 = V$), correspondingly. For the same size of the electrodes and the same ionic solution, the resistance of the solution is proportional to the distance between the electrodes. As a result, the smaller cells, for which the distance between the electrodes has been reduced, are characterised by a smaller electrical resistance. According to inequality (8.10), for $n = 2$, the next inequality holds:

$$\frac{V_1}{r_1} + \frac{V_2}{r_2} \geq 2\frac{V_1 + V_2}{r_1 + r_2} = \frac{2V}{R} \tag{8.16}$$

Multiplying both sides of inequality (8.16) with the positive value $Z \times t$ gives

$$Z\left(V_1 / r_1\right)t + Z\left(V_2 / r_2\right)t \geq 2Z\left(V / R\right)t \tag{8.17}$$

The left hand side of inequality (8.17) can be physically interpreted as the sum of masses $m_1 + m_2$ in grams, of substance deposited at the electrodes of the smaller electrolytic cells, by currents of magnitudes $I_1 = V_1/r_1$ and $I_2 = V_2/r_2$, for a duration of t seconds. The right-hand side of inequality (8.17) can be physically interpreted as the mass m_0 of substance deposited at the electrode of the original cell by current of magnitude $I = V/R$, for the same duration of t seconds. Inequality (8.17) predicts that as a result of the cell's segmentation, the mass of deposited substance can be increased more than twice.

Again, the presence of asymmetry in inequality (8.17) is a condition for improved performance. For $V_1/r_1 = V_2/r_2$, no increase of the deposited substance is present.

The segmentation of additive quantities through algebraic inequality (8.10) can be used to achieve systems and processes with superior performance and the algebraic inequality is applicable in any area of science and technology, as long as the additive quantities comply with the simple condition (8.11).

Inequality (8.10) can also be rewritten as

$$\frac{1}{n}\left(\frac{a_1}{x_1} + \frac{a_2}{x_2} + \ldots + \frac{a_n}{x_n}\right) \geq \frac{a_1 + a_2 + \ldots + a_n}{x_1 + x_2 + \ldots + x_n} \tag{8.18}$$

Even if the ratios a_i/x_i are not additive quantities, the left-hand side of the inequality can still be interpreted as an estimate of the intensive quantity $\dfrac{a_1 + a_2 + \ldots + a_n}{x_1 + x_2 + \ldots + x_n}$ through the average of the segmented intensive quantities a_i/x_i. Here is an example.

Suppose that a mixture of n distinct incompressible substances has been specified through the masses m_i of the substances and the volumes v_i these substances occupy. The density of the mixture is clearly $\rho_{\text{mix}} = \dfrac{m_1 + m_2 \ldots + m_n}{v_1 + v_2 + \ldots + v_n}$. However, if the density of the mixture is estimated by using $\hat{\rho}_{\text{mix}} = \dfrac{1}{n}\left(\dfrac{m_1}{v_1} + \dfrac{m_2}{v_2} + \ldots + \dfrac{m_n}{v_n}\right)$, this could result in an overestimation of the real density.

The overestimation can be significant which will be illustrated by the next numerical example. For three distinct substances with masses $m_1 = 125g, m_2 = 50g$

and $m_3 = 25g$ and volumes $v_1 = 10cm^3$, $v_2 = 15cm^3$ and $v_3 = 20cm^3$, correspondingly, inequality (8.18) becomes

$$\frac{1}{3}\left(\frac{m_1}{v_1} + \frac{m_2}{v_2} + \frac{m_3}{v_3}\right) \geq \frac{M}{V} \qquad (8.19)$$

The average density estimated by the left side of inequality (8.19) is

$(1/3)(m_1/v_1 + m_2/v_2 + m_3/v_3) = (1/3)[125/10 + 50/15 + 25/20] = 5.69\,g/cm^3$.

This estimated value is significantly larger than the correct value $M/V = 200/45 = 4.44\,g/cm^3$ for the density of the mixture.

8.2.4   DEFLECTION OF ELASTIC ELEMENTS CONNECTED IN SERIES AND PARALLEL

If the variables a_i in inequality (8.10) are interpreted as forces (loads) and x_i are interpreted as stiffness values of elastic elements (springs), the terms a_i/x_i can be interpreted as deflections of the springs under these loads (Todinov, 2022b). The forces a_i are additive quantities and the stiffness values for elastic elements connected in parallel are also additive quantities.

Suppose also that $a_1 = a_2 = \ldots = a_n = F/n$. If x_i are ordered in ascending order $(x_1 \leq x_2 \leq \ldots \leq x_n)$, because $a_1 = a_2 = \ldots = a_n = F/n$, the conditions $a_1 \geq a_2 \geq \ldots \geq a_n$, $x_1 \leq x_2 \leq \ldots \leq x_n$ for the validity of inequality (8.10) are fulfilled. Inequality (8.10) can then be rewritten as

$$\frac{F}{x_1 + x_2 + \ldots + x_n} \leq \frac{1}{n}\left(\frac{F/n}{x_1} + \frac{F/n}{x_2} + \ldots + \frac{F/n}{x_n}\right) = \frac{1}{n^2}\left(\frac{F}{x_1} + \frac{F}{x_2} + \ldots + \frac{F}{x_n}\right) \quad (8.20)$$

The left-hand side of inequality (8.20) can be physically interpreted as deflection of an assembly of n elastic elements connected in parallel, with stiffnesses x_i (see Figure 8.2a) and loaded with a force with magnitude F. The expression $F/x_1 + \ldots + F/x_n$ in the right-hand side of inequality (8.20), is the total deflection of the same n elastic elements connected in series and loaded with a single force of the same magnitude F (Figure 8.2b).

The physical interpretation of inequality (8.20) states that the deflection of n elastic elements connected in parallel, is more than n^2 times smaller than the deflection of the same elastic elements connected in series and loaded with a force of the same magnitude.

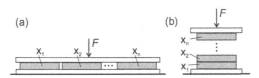

FIGURE 8.2   Elastic elements with stiffness values x_i arranged in (a) parallel and (b) series. The series arrangement is loaded with the same force F as the parallel arrangement.

8.2.5   OPTIMAL ALLOCATION OF RESOURCES FOR MAXIMISING EXPECTED BENEFIT

Reverse engineering of inequality (8.6) can also be applied for optimal resource allocation.

Suppose that a certain amount p of resources (for example, $p = 9$ research units) are available in a particular field which includes q different topics (for example, $q = 6$) that must be explored experimentally. From past experience, it is known that the expected number of results E obtained from the experimental exploration is proportional to the density p/q of the resources engaged in the selected set of topics, but does not depend on the range of explored topics: $E = \lambda p/q$. The numerical value of the constant of proportionality λ is unknown.

This statement of the optimisation problem fits in the description of the model presented with inequality (8.6). In the absence of segmentation, the expected number of results is, $E = \lambda p/q = \lambda \times 9/6 = 1.5\lambda$ where λ is unknown coefficient of proportionality.

Suppose that the available resources are segmented into $p_1 = 6$ and $p_2 = 3$ research units and the available topics to be explored into $q_1 = 2$ and $q_2 = 4$. The first resource segment can be paired with the first segment of research topics while the second resource segment can be paired with the second segment of research topics such that the inequalities $p_1 > p_2$ and $q_1 < q_2$ are fulfilled. In this case, inequality (8.6) holds true. The inequality predicts at least a two-fold increase in the expected number of results ($n = 2$). Indeed, segmentation of the resources and research topics yields

$$E_1 + E_2 = \left(\frac{\lambda p_1}{q_1} + \frac{\lambda p_2}{q_2} \right) = \lambda \times (6/2 + 3/4) = 3.75\lambda$$

number of results while the lack of segmentation yields

$$E = \lambda p / q = \lambda \times 9/6 = 1.5\lambda$$

number of results.

Since, $(E_1 + E_2)/E = 2.5$, there is a 2.5 times increase in the expected number of results as a result of the segmentation. In inequality (8.6), the constant λ is cancelled and its specific numerical value does not affect the predictions of the inequality regarding the effect of the segmentation.

Here, it is important to note that not every segmentation results in more than a two-fold increase in the expected number of results. For example, a more than a two-fold increase in the expected results will not occur if the available resources are segmented into $p_1 = 6$, $p_2 = 3$ and the topics to be explored experimentally are divided into $q_1 = 4$ and $q_2 = 2$ topics. Indeed, in this case, $E = \lambda p/q = \lambda \times 9/6 = 1.5\lambda$ and

$$E_1 + E_2 = \left(\frac{\lambda p_1}{q_1} + \frac{\lambda p_2}{q_2} \right) = \lambda (6/4 + 3/2) = 3\lambda = 2 \times (1.5\lambda)$$

9 Optimal Selection and Expected Time of Unsatisfied Demand by Reverse Engineering of Algebraic Inequalities

9.1 MAXIMISING THE PROBABILITY OF SUCCESSFUL SELECTION FROM SUPPLIERS WITH UNKNOWN PROPORTIONS OF RELIABLE COMPONENTS

Consider the non-trivial abstract inequalities:

$$x_1^2 + x_2^2 + x_3^2 \geq x_1 x_2 + x_2 x_3 + x_3 x_1 \tag{9.1}$$

$$x_1^3 + x_2^3 + x_3^3 \geq 3 x_1 x_2 x_3 \tag{9.2}$$

$$2\left(x_1^3 + x_2^3 + x_3^3\right) \geq x_1^2 x_2 + x_1^2 x_3 + x_2^2 x_1 + x_2^2 x_3 + x_3^2 x_1 + x_3^2 x_2 \tag{9.3}$$

$$2\left(x_1^4 + x_2^4 + x_3^4\right) \geq x_1^3 x_2 + x_1^3 x_3 + x_2^3 x_1 + x_2^3 x_3 + x_3^3 x_1 + x_3^3 x_2 \tag{9.4}$$

Inequalities (9.1)–(9.4) can be proved by invoking the Muirhead's inequality introduced in Chapter 2. For any set of non-negative numbers $x_1, x_2, \ldots, x_n$, a symmetric sum is defined as $\sum_{\text{sym}} x_1^{a_1} x_2^{a_2} \ldots x_n^{a_n}$ which, when expanded, includes $n!$ terms. Each term is formed by a distinct permutation of the sequence $a_1, a_2, \ldots, a_n$.

The Muirhead's inequality introduced in Section 2.1.7 states that if the sequence $\{a\}$ is majorizing sequence $\{b\}$ and $x_1, x_2, \ldots, x_n$ are non-negative, the next inequality holds:

$$\sum_{\text{sym}} x_1^{a_1} x_2^{a_2} \ldots x_n^{a_n} \geq \sum_{\text{sym}} x_1^{b_1} x_2^{b_2} \ldots x_n^{b_n} \tag{9.5}$$

DOI: 10.1201/9781003517764-9

Consider the two non-increasing sequences $a_1 \geq a_2 \geq, ..., \geq a_n$ and $b_1 \geq b_2 \geq, ..., \geq b_n$ of non-negative real numbers. The sequence $\{a\}$ is said to majorize sequence $\{b\}$ if the following conditions are fulfilled:

$$a_1 \geq b_1; a_1 + a_2 \geq b_1 + b_2; ...; a_1 + a_2 + ... + a_{n-1} \geq b_1 + b_2 + ... + b_{n-1};$$

$$a_1 + a_2 + ... + a_{n-1} + a_n = b_1 + b_2 + ... + b_{n-1} + b_n \qquad (9.6)$$

Consider now the sequences $\{a\} = [2, 0, 0]$ and $\{b\} = [1, 1, 0]$. Because the sequence $\{a\}$ majorizes sequence $\{b\}$, inequality (9.7) is obtained.

$$2! \times \left(x_1^2 + x_2^2 + x_3^2 \right) \geq 2x_1x_2 + 2x_2x_3 + 2x_3x_1 \qquad (9.7)$$

Dividing both sides of (9.7) by 2! transforms inequality (9.7) into inequality (9.1).

Consider now the sequences $\{a\} = [3, 0, 0]$ and $\{b\} = [1, 1, 1]$. Because the sequence $\{a\}$ majorizes sequence $\{b\}$, inequality (9.8) is obtained.

$$2! \times \left(x_1^3 + x_2^3 + x_3^3 \right) \geq 3! \times x_1x_2x_3 \qquad (9.8)$$

Dividing both sides of (9.8) by 2! transforms inequality (9.8) into inequality (9.2).

Next, since the sequence $\{a\} = [3, 0, 0]$ majorizes the sequence $\{b\} = [2, 1, 0]$, the following inequality also follows immediately from the Muirhead's inequality (9.5):

$$2! \times \left(x_1^3 + x_2^3 + x_3^3 \right) \geq x_1^2 x_2 + x_1^2 x_3 + x_2^2 x_1 + x_2^2 x_3 + x_3^2 x_1 + x_3^2 x_2 \qquad (9.9)$$

which gives inequality (9.3).

Finally, since the sequence $\{a\} = [4, 0, 0]$ majorizes the sequence $\{b\} = [3, 1, 0]$, the following inequality follows immediately from the Muirhead's inequality (9.5):

$$2! \left(x_1^4 + x_2^4 + x_3^4 \right) \geq x_1^3 x_2 + x_1^3 x_3 + x_2^3 x_1 + x_2^3 x_3 + x_3^3 x_1 + x_3^3 x_2 \qquad (9.10)$$

which is effectively inequality (9.4).

Inequalities (9.1)–(9.4) can be reverse-engineered in a natural way if the constraints $0 \leq x_1 \leq 1$, $0 \leq x_2 \leq 1$ and $0 \leq x_3 \leq 1$ are imposed on the variables entering the inequalities. If the left- and the right-hand side of inequality (9.1) are multiplied by 1/3, the inequality

$$(1/3)x_1^2 + (1/3)x_2^2 + (1/3)x_3^2 \geq (1/3)x_1x_2 + (1/3)x_2x_3 + (1/3)x_3x_1 \qquad (9.11)$$

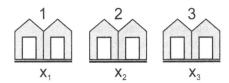

FIGURE 9.1   Three suppliers delivering reliable components with unknown proportions x_1, x_2 and x_3.

is obtained, whose left- and right-hand side can be physically interpreted as probabilities of mutually exclusive events. The factors 1/3 on the left-hand side of (9.11) can be interpreted as probabilities of random selection of a particular supplier from three available suppliers (Figure 9.1). Each supplier offers components of the same type and the variables x_1, x_2 and x_3 can be interpreted as the fractions of reliable components characterising the individual suppliers ($0 \le x_i \le 1$), correspondingly. It is important to emphasize that the fractions x_i of reliable components characterising the individual suppliers *are unknown*.

For example, if it comes to suppliers of suspension automotive springs, reliable components are those which can endure more than a specified number of cycles when tested on a specially designed test rig. The remaining springs fail significantly below the specified number of cycles and are considered to be of inferior reliability.

The left-hand side of inequality (9.11) can be physically interpreted as the probability of purchasing two reliable components from a randomly selected supplier.

Indeed, purchasing two reliable components from the same supplier can occur in three mutually exclusive ways: (i) the first supplier is randomly selected and both components purchased from the first supplier are reliable, the probability of which is $(1/3)x_1^2$; (ii) the second supplier is randomly selected and both components purchased from the second supplier are reliable, the probability of which is $(1/3)x_2^2$; and finally, the third supplier is randomly selected and both components purchased from the third supplier are reliable, the probability of which is $(1/3)x_3^2$. According to the total probability theorem, the probability of occurrence of any of these three mutually exclusive events is

$$p_1 = (1/3)x_1^2 + (1/3)x_2^2 + (1/3)x_3^2$$

which is the left-hand side of inequality (9.11). It is assumed that the number of components offered by each supplier is sufficiently large so that the probability of purchasing a second reliable component is practically equal to the probability that the first purchased component will be reliable.

The right-hand side of inequality (9.11) can be physically interpreted as the probability of purchasing two reliable components from two different, randomly selected suppliers. Indeed, purchasing two reliable components from two suppliers can occur in three mutually exclusive ways: (i) suppliers 1 and 2 are randomly selected and both components purchased from these suppliers are reliable (with probability $(1/3)x_1x_2$); (ii) suppliers 2 and 3 are randomly selected, and both components purchased from these suppliers are reliable (with probability $(1/3)x_2x_3$); and finally, suppliers 3 and 1 are randomly selected and the components purchased from these suppliers are reliable (with probability $(1/3)x_3x_1$). According to the total probability theorem, the probability of occurrence of any of these three mutually exclusive events is

$$p_2 = (1/3)x_1x_2 + (1/3)x_2x_3 + (1/3)x_3x_1$$

which is the right-hand side of inequality (9.11).

The prediction of inequality (9.11) states that purchasing both components from a single, randomly selected supplier, is characterised by a higher probability of purchasing components which are all reliable (Todinov, 2020d). This is *a surprising and highly counter-intuitive result*. After all, the proportions x_i of reliable components characterising the suppliers are unknown.

The difference in the probabilities p_1 and p_2 evaluated from the left- and right-hand side of inequality (9.11) can be significant. Thus for $x_1 = 0.5$, $x_2 = 0.9$ and $x_3 = 0.1$, the left-hand side of inequality (9.11) gives:

$$p_1 = (1/3) \times 0.5^2 + (1/3) \times 0.9^2 + (1/3) \times 0.1^2 = 0.357$$

while the right-hand side gives

$$p_2 = (1/3) \times 0.5 \times 0.9 + (1/3) \times 0.9 \times 0.1 + (1/3) \times 0.1 \times 0.5 = 0.197$$

Equality in (9.11) is attained only for $x_1 = x_2 = x_3$.

It must be pointed out that selecting two components from a single, randomly selected supplier will also maximise the probability that both components will be unreliable.

This, however, does not mean that it is more beneficial to select the components from two different suppliers. Despite that selecting from two different suppliers decreases the probability that both selected components will be unreliable, for an unreliable system to be present it is not necessary both selected components to be unreliable. Selecting a single unreliable component is sufficient. Selecting components from two different suppliers decreases the probability of having a reliable system (two reliable components). This conclusion remains unchanged if the fractions of reliable components are considered to be fractions of faulty components.

Indeed, if we consider $a = 0.85$, $b = 0.24$ and $c = 0.57$, to be the fractions of unreliable components in the batches, the probability of selecting two reliable components from a randomly selected batch is

$$p_1 = (1/3)\left[(1-a)^2 + (1-b)^2 + (1-c)^2\right] = (1/3)\left[0.15^2 + 0.76^2 + 0.43^2\right] = 0.262$$

while the probability of selecting two reliable components from two randomly selected batches is

$$p_2 = (1/3)\big[(1-a)(1-b)+(1-b)(1-c)+(1-c)(1-a)\big]$$
$$= (1/3)\big[0.15\times0.76+0.76\times0.43+0.43\times0.15\big] = 0.168$$

Again, $p_1 > p_2$.

No matter what a, b and c denote (percentage of reliable components or percentage of unreliable components) the probability that both components will be reliable is always maximised by selecting both components from the same randomly selected batch. With this, the probability of building a reliable system is also maximised.

In a similar fashion, by using the Muirhead's inequality, inequality (9.5) can be generalised for more than two selected components. (The reasoning is very similar to the reasoning in deriving inequalities (9.1)–(9.4) and will not be repeated).

Similar interpretation can be made for inequality (9.2). If the left- and right-hand side of inequality (9.2) are multiplied by 1/3, the inequality

$$(1/3)x_1^3 + (1/3)x_2^3 + (1/3)x_3^3 \ge x_1x_2x_3 \tag{9.12}$$

is obtained. The left-hand side of inequality (9.12) can be physically interpreted as the probability of purchasing three reliable components from a randomly selected supplier. The right-hand side of inequality (9.12) can be physically interpreted as the probability of purchasing three reliable components from the three available suppliers.

Again, inequality (9.12) predicts that if the proportions x_i of reliable components characterising the suppliers are unknown, purchasing the three components from a single, randomly selected supplier, will maximize the probability that all purchased components will be reliable.

Next, consider inequality (9.3). If the left and the right-hand side of inequality (9.3) are multiplied by 1/6, the inequality

$$(1/3)x_1^3 + (1/3)x_2^3 + (1/3)x_3^3 \ge (1/6)x_1^2x_2 + (1/6)x_1^2x_3$$
$$+ (1/6)x_2^2x_1 + (1/6)x_2^2x_3 + (1/6)x_3^2x_1 + (1/6)x_3^2x_2 \tag{9.13}$$

is obtained.

The left-hand side of inequality (9.13) is the probability of purchasing three reliable components from a randomly selected supplier. The right-hand side of inequality (9.13) is the probability of purchasing three reliable components from two randomly selected suppliers. Again, inequality (9.13) predicts that purchasing all components from a single, randomly selected supplier is associated with a higher probability of purchasing three reliable components.

Finally, consider inequality (9.4). If the left and the right-hand side of inequality (9.4) are multiplied by 1/6, the inequality

$$
\begin{aligned}
&(1/3)x_1^4 +(1/3)x_2^4 +(1/3)x_3^4 \geq (1/6)x_1^3 x_2 +(1/6)x_1^3 x_3 \\
&+(1/6)x_2^3 x_1 +(1/6)x_2^3 x_3 +(1/6)x_3^3 x_1 +(1/6)x_3^3 x_2
\end{aligned}
\tag{9.14}
$$

is obtained.

The left-hand side of inequality (9.14) is the probability of purchasing four reliable components from a randomly selected supplier. The right-hand side of inequality (9.14) is the probability of purchasing four reliable components from two suppliers by purchasing 3 components from one randomly selected supplier and the other component from another randomly selected supplier.

The generalization of these results for different numbers of components and different numbers of suppliers leads to the following conclusion related to suppliers delivering the same type of components and characterized by unknown fractions of reliable components. If no information is available about the fractions of reliable components characterizing the individual suppliers, the best strategy for purchasing only reliable components is to purchase all components from a single, randomly selected supplier. In spite of the complete lack of knowledge related to the proportions of reliable components characterizing the separate suppliers and despite existing dependencies among the suppliers, purchasing all components from a single, randomly selected supplier is characterized by the highest probability that all purchased components will be reliable.

These highly counter-intuitive results contradict the conventional habit of advocating diversification as a risk reduction measure and expose the dangers of blindly following conventional wisdom in risk reduction.

The conclusions in this section have been confirmed by Monte Carlo simulations (Todinov, 2020d).

9.2 INCREASING THE PROBABILITY OF SUCCESSFUL ACCOMPLISHMENT OF TASKS BY DEVICES WITH UNKNOWN RELIABILITY

The inequalities from the previous section have alternative useful physical interpretations. Consider for example, inequalities (9.12) and (9.13), which, for convenience, will be presented in the form:

$$
\frac{1}{3}x^3 +\frac{1}{3}y^3 +\frac{1}{3}z^3 \geq xyz
\tag{9.15}
$$

$$
\begin{aligned}
&(1/3)x^3 +(1/3)y^3 +(1/3)z^3 \geq (1/6)x^2 y +(1/6)x^2 z +(1/6)y^2 x \\
&+(1/6)y^2 z +(1/6)z^2 x +(1/6)z^2 y
\end{aligned}
\tag{9.16}
$$

where $0 \leq x, y, z \leq 1$.

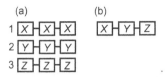

FIGURE 9.2 (a) An arrangement of three devices of the same type allocated to three identical tasks; (b) an arrangement of three devices of different type assigned to three identical tasks.

The factor $1/3$ on the left-hand side of inequality (9.15) can be physically interpreted as the probability of selecting any of the three arrangements (Figure 9.2a, arrangements 1, 2, and 3). Each arrangement includes three devices of types X, Y, and Z assigned to a mission comprising three identical tasks. The probabilities x, y, and z of successful accomplishment of a task by a device of type X, Y, and Z are unknown.

A mission is considered to have been accomplished successfully if all three tasks assigned to the devices have been accomplished successfully. According to the total probability theorem, the left side of inequality (9.15) represents the total probability that a randomly selected arrangement consisting of three devices of the same type, allocated to three identical tasks (Figure 9.2a), will successfully accomplish a mission.

Indeed, a successful accomplishment of a mission by devices of the same type can occur in three different, mutually exclusive ways: (i) arrangement 1, including devices of type X is randomly selected, and all devices successfully accomplish their tasks, the probability of which is $(1/3)x^3$; (ii) arrangement 2 including devices of type Y is randomly selected, and all devices successfully accomplish their tasks, the probability of which is $(1/3)y^3$ and finally, (iii) arrangement 3 including devices of type Z is randomly selected and all devices successfully accomplish their tasks, the probability of which is $(1/3)z^3$.

The right-hand side of inequality (9.15) is the probability xyz that the devices of different types $(X, Y$ and $Z)$ successfully accomplish their tasks.

The separate terms of inequality (9.16) can also be physically interpreted by using the total probability theorem for mutually exclusive events. According to the total probability theorem, the left-hand side of inequality (9.16) represents the total probability that a randomly selected arrangement consisting of three identical type devices will successfully accomplish the mission (Figure 9.3a).

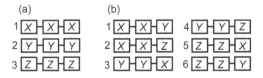

FIGURE 9.3 (a) Arrangements of three devices of the same type assigned to three identical tasks; (b) arrangements of three devices two of which are of the same type, assigned to three identical tasks.

According to the total probability theorem, the right-hand side of inequality (9.16) represents the total probability that a randomly selected arrangement composed of three devices, two of which are of the same type, will successfully accomplish the mission. A successful accomplishment of a mission including three devices, two of which are of the same type, can occur in 6 different ways (Figure 9.3b): (i) the device arrangement (X,X,Y) is randomly selected, and all three tasks are successfully accomplished by the devices, the probability of which is $(1/6)x^2y$; (ii) the device arrangement (X,X,Z) is randomly selected, and all three tasks are successfully accomplished by the devices, the probability of which is $(1/6)x^2z$ and so on.

Although the probabilities x, y and z of successful accomplishment of a task by the devices of different types are *unknown*, according to inequalities (9.15) and (9.16), the strategy of randomly selecting an arrangement composed of three devices of the same type is characterised by a higher chance of a successful accomplishment of the mission compared to selecting an arrangement composed of three different types of devices or selecting an arrangement composed of three devices two of which are of the same type (Todinov, 2022c).

This result is counterintuitive given that the probabilities x,y and z of successfully completing a task by the different types of devices are unknown. Why should any particular strategy have an advantage if these probabilities are unknown? It seems that in a situation of profound uncertainty, no specific strategy should matter. However, despite the profound uncertainty regarding the probabilities of task completion by the devices, selecting a random arrangement consisting of the same type of device is always the best strategy. This conclusion holds true regardless of any unknown interdependencies among the probabilities of successful accomplishment of the tasks by the different devices.

The significant advantage provided by the superior strategy can be illustrated by a numerical example. Suppose that the probabilities of accomplishing the tasks characterising the different device types X, Y and Z are $x = 0.88$, $y = 0.64$ and $z = 0.38$. According to inequality (9.15), the probability that a randomly selected arrangement including three devices of the same type will successfully accomplish the mission is

$$p_1 = (1/3) \times 0.88^3 + (1/3) \times 0.64^3 + (1/3) \times 0.38^3 = 0.33$$

while the probability of accomplishing successfully the mission by three devices of different types is $p_2 = 0.88 \times 0.64 \times 0.38 = 0.21$.

According to inequality (9.16), the probability of accomplishing the mission by a randomly selected arrangement consisting of three devices, two of which are of the same type, is:

$$p_3 = (1/6) \times [0.88^2 \times 0.64 + 0.88^2 \times 0.38 + 0.64^2 \times 0.88 + 0.64^2 \times 0.38$$
$$+ 0.38^2 \times 0.88 + 0.38^2 \times 0.64] = 0.25$$

By using Muirhead's inequality for $n > 3$, these results are naturally generalized for $n > 3$ tasks composing the mission (the generalization is straightforward, and details have been omitted to conserve space). The strategy of randomly selecting an arrangement composed of n devices of the same type is characterized by the highest chance of successfully accomplishing the mission compared to any other arrangement.

9.3 MONTE CARLO SIMULATIONS

The predictions from inequalities (9.11)–(9.14) have been confirmed by Monte-Carlo simulations, each of which involved 10 million trials. For fractions of reliable components of 0.9, 0.55 and 0.35, characterising the three suppliers, the Monte-Carlo simulation resulted in probabilities $p_1 = 0.41$ and $p_2 = 0.33$ of purchasing two reliable components from a randomly selected single supplier and from two randomly selected suppliers, correspondingly. The left and right part of the inequality (9.11) give $p_1 = 0.41$ and $p_2 = 0.33$ for the same probabilities, which confirms the validity of inequality (9.11).

For fractions of reliable components characterising the three suppliers $x_1 = 0.9$, $x_2 = 0.75$ and $x_3 = 0.25$, the Monte-Carlo simulation based on 10 million trials resulted in probabilities $p_1 = 0.389$ and $p_2 = 0.26$ and $p_3 = 0.169$ of purchasing three reliable components from a randomly selected single supplier, from two randomly selected suppliers and from all three available suppliers, correspondingly. The left- and right-hand sides of inequalities (9.13) and (9.12) yield $p_1 = 0.389$, $p_2 = 0.26$ and $p_3 = 0.169$ for the same probabilities, which confirms the validity of inequalities (9.15) and (9.16).

For reliability fractions of 0.9, 0.4 and 0.3 characterising the suppliers, the Monte Carlo simulations yielded the value 0.23 for the probability of 4 reliable components purchased from a randomly selected supplier and the value 0.1 for the probability of 4 reliable components if 3 components are purchased from a randomly selected supplier and the remaining component from another randomly selected supplier. These results are confirmed by the probabilities calculated from the left- and right-hand side of inequality (9.14).

9.4 ASSESSING THE EXPECTED TIME OF UNSATISFIED DEMAND FROM USERS PLACING RANDOM DEMANDS IN A TIME INTERVAL

Complex algebraic inequalities can also be reverse engineered successfully. Consider the complex algebraic inequality

$$(1-\psi)^n + n\psi^1(1-\psi)^{n-1} + \frac{n(n-1)}{1 \times 2}\psi^2(1-\psi)^{n-2} + \dots$$
$$+ \frac{n(n-1)\dots(n-m+1)}{1 \times 2 \times \dots \times m}\psi^m(1-\psi)^{n-m} \le 1 \qquad (9.17)$$

where $0 \le \psi \le 1$, n and m are integers for which $m \le n$; $n > 1$.

Inequality (9.17) can be proved easily by observing that it has been obtained from the binomial expansion of the expression

$$\left[\left(1-\psi\right)+\psi\right]^{n}=1 \tag{9.18}$$

and discarding from the expansion all positive terms corresponding to $m+1$, $m+2$, ..., n. Equality in (9.17) is attained for $\psi = 0$ or $m = n$.

Inequality (9.17) can be reverse engineered if n stands for the number of users placing random demands for a particular resource over the time interval $(0,L)$ and ψ stands for the time fraction of random demand over the time interval $(0,L)$. For example, if the duration of each random demand is d, $\psi = d/L$ is the time fraction of random demand.

Consider users demanding randomly a particular single resource (piece of equipment, repairers, procedure of a certain duration, etc.), during an operation period with length L. The available resource (e.g. piece of equipment) can satisfy only a single random demand at a time. The start of random demand i is marked by s_i and the end of demand i is marked by e_i. Figure 9.4 depicts three random demands with durations of length d appearing at random times s_1, s_2 and s_3. The overlapping region $s_3 e_2$ in the figure marks the simultaneous existence of two random demands which leads to unsatisfied demand.

For complex systems (production system, computer network, etc.), the random demands can be demands for repair from failed components building the system. The random demands can also be demands for a particular piece of life-saving equipment from critically ill patients, etc.

The first term $(1-\psi)^n$ of inequality (9.17) is the expected fraction of time during which no random demand is present. Indeed, the expected fraction of time during which no random demand is present is equal to the probability that a randomly selected point in the time interval $(0,L)$ will not be 'covered' by any of the n randomly placed demands (segments with length d).

The second term $n\psi^1(1-\psi)^{n-1}$ of inequality (9.17) is the expected fraction of time during which exactly one random demand is present in the time interval $(0,L)$. Indeed this expected time fraction is equal to the probability that a randomly selected point in the time interval $(0,L)$ will be covered by exactly one of the n randomly placed demands (segments with length d). This probability is a sum of the probabilities $\psi^1(1-\psi)^{n-1}$ of n mutually exclusive events: exactly one segments of length d covers the selected point and none of the other segments covers it.

FIGURE 9.4   Random demands for a particular resource over a time interval $0,L$.

The third term $\dfrac{n(n-1)}{1\times 2}\psi^2(1-\psi)^{n-2}$ of inequality (9.17) is the expected fraction of time during which exactly two random demands are present. This time fraction is equal to the probability that a randomly selected point in the time interval $(0,L)$ will be covered by exactly two of the n randomly placed segments with length d. This is a sum of the probabilities $\psi^2(1-\psi)^{n-2}$ of $\dfrac{n(n-1)}{1\times 2}$ mutually exclusive events: exactly two segments of length d cover the selected point and none of the other segments covers it.

The $m+1$st term $\dfrac{n(n-1)...(n-m+1)}{1\times 2\times ...\times m}\psi^m(1-\psi)^{n-m}$ of inequality (9.17) is the expected fraction of time during which exactly m random demands are present. This time fraction is equal to the probability that a randomly selected point in the time interval $(0,L)$ will be covered by exactly m of the n randomly placed segments with length d. It is a sum of the probabilities $\psi^m(1-\psi)^{n-m}$ of $\dfrac{n(n-1)...(n-m+1)}{1\times 2\times ...\times m}$ mutually exclusive events: exactly m segments, representing the random demands from m users, cover the selected point and none of the other segments covers it. The sum of the expected fractions of time representing $0, 1, 2,...,n$ simultaneous random demands is equal to one because $[(1-\psi)+\psi]^n = 1$.

Because the sum of all expected time fractions corresponding to $0, 1, 2,...,n$ simultaneously present random demands is equal to one, inequality (9.17) effectively states that the sum of the expected fractions of time during which there is no random demand, there is exactly one random demand, exactly two random demands,..., exactly m random demands ($m \le n$), cannot exceed unity.

Suppose that the available resources can satisfy m (or fewer) simultaneously present random demands but not more than m random demands. Because the sum of all expected time fractions corresponding to $0, 1, 2,...,n$ simultaneous random demands is equal to one, the total expected fraction of time during which there is unsatisfied demand ($m+1$ or more random demands are simultaneously present) is given by

$$p = 1 - \left((1-\psi)^n + n(1-\psi)^{n-1}\psi^1 + \frac{n(n-1)}{1\times 2}(1-\psi)^{n-2}\psi^2 + ... \right.$$
$$\left. + \frac{n(n-1)...(n-m+1)}{1\times 2\times ...\times m}(1-\psi)^{n-m}\psi^m \right) \ge 0$$

This inequality is effectively the rearranged inequality (9.17) and p is the probability of unsatisfied demand due to clustering of random demands in the time interval $(0,L)$.

Clearly, the compact inequality (9.17) describes the behaviour of a complex system with a number of features: (i) an arbitrary number n of users initiating demands, (ii) each demand is randomly placed along a specified time interval, (iii) different number of sources available for servicing the random demands.

It can be shown that inequality (9.17) is valid also if the demand time is not fixed but varies, with mean equal to d (Todinov, 2017).

9.4.1 APPLICATIONS OF INEQUALITY (9.17)

Inequality (9.17) has a wide range of applications.

- The competition of random demands for a particular single resource/service on a finite time interval is a common example of risk controlled by the simultaneous presence of critical events. The appearance of a critical event engages the servicing resource and if a new critical event occurs during the service time of the first critical event, no servicing resource will be available for the second event.

Suppose that only a single repair unit is available for servicing failures on a power distribution system. If a power line failure occurs, the repair resource will be engaged and if another failure occurs during the repair time d associated with the first failure, no free repair resource will be available to recover from the second failure. The delay in the second repair could lead to overloading of the power distribution system, thereby inducing further failures.

There are cases where the probability of a simultaneous presence (overlapping) of risk-critical critical events must be low. A low probability of a simultaneous presence of random demands is required in situations where people in critical condition demand a particular piece of life-saving equipment for time interval with length d. If only a single piece of life-saving equipment is available, simultaneous demands within a time interval with length d cannot be satisfied and the consequences could be fatal. Here are other applications of inequality (9.17).

- *Stored spare equipment in a warehouse servicing the needs of customers arriving randomly during a specified time interval*:
  After a demand from a customer, the warehouse requires a minimum period of length d to restore/return the dispatched equipment before the next demand can be serviced. In this case, the probability of unsatisfied demand equals the probability of two or more customer arrivals clustering within the critical period needed to make the equipment available for the next customer.
- *Supply systems that accumulate the supplied resource before it is dispatched for consumption (compressed gaseous substances, for example)*.
  Suppose that, after a demand for the resource, the system needs a minimum period of average duration d to restore the amount of supplied resource to the level existing before the demand. In this case, the probability of unsatisfied demand equals the probability of two or more random demands clustering within the critical recovery period d.

- *Appearance of a shock event in a time interval.*
  A related risk controlled by overlapping (simultaneous presence) of critical events is the appearance of a shock that requires a particular minimum time interval for the system to recover. If another critical event appears before the system has recovered, the system's strength or capacity is exceeded, resulting in system failure. Consider, for example, shocks caused by failures associated with pollution to the environment (e.g., a leakage of chemicals). A failure followed by another failure associated with leakage of chemicals before a critical recovery time interval has elapsed could result in irreparable damage to the environment. For instance, clustering of failures associated with the release of chemicals in the seawater could result in a dangerously high acidity, which would destroy marine life.

In all of the listed examples, the overlapping of the risk-critical random events cannot be avoided; therefore, the level of risk correlates with the expected time fraction of overlapping risk-critical events.

10 Enhancing Systems and Process Performance by Reverse Engineering of Algebraic Inequalities Based on Sub-Additive and Super-Additive Functions

10.1 INCREASING THE ABSORBED KINETIC ENERGY DURING A PERFECTLY INELASTIC COLLISION

Another special case of the general sub-additive inequality (2.38) is the algebraic inequality

$$\frac{a_1^2}{a_1+b_1} + \frac{a_2^2}{a_2+b_2} + \ldots + \frac{a_n^2}{a_n+b_n} \geq \frac{a^2}{a+b} \tag{10.1}$$

where both controlling factors $a = a_1 + a_2 + \ldots + a_n$ and $b = b_1 + b_2 + \ldots + b_n$ are additive positive quantities. For $n = 2$, inequality (10.1) becomes

$$\frac{a_1^2}{a_1+b_1} + \frac{a_2^2}{a_2+b_2} \geq \frac{\left(a_1+a_2\right)^2}{a_1+b_1+a_2+b_2} \tag{10.2}$$

The special case (10.2) can be proved by showing that $\dfrac{a_1^2}{a_1+b_1} + \dfrac{a_2^2}{a_2+b_2} - \dfrac{\left(a_1+a_2\right)^2}{a_1+b_1+a_2+b_2} \geq 0$. Since

$$\frac{a_1^2}{a_1+b_1} + \frac{a_2^2}{a_2+b_2} - \frac{\left(a_1+a_2\right)^2}{a_1+b_1+a_2+b_2} = \frac{\left(a_1b_2 - a_2b_1\right)^2}{\left(a_1+b_1\right)\left(a_2+b_2\right)\left(a_1+b_1+a_2+b_2\right)} \tag{10.3}$$

is a positive number, this proves the special case (10.2).

DOI: 10.1201/9781003517764-10

Adding the term $\dfrac{a_3^2}{a_3 + b_3}$ to both sides of the already proved inequality (10.2) results in

$$\frac{a_1^2}{a_1 + b_1} + \frac{a_2^2}{a_2 + b_2} + \frac{a_3^2}{a_3 + b_3} \geq \frac{\left(a_1 + a_2\right)^2}{a_1 + a_2 + b_1 + b_2} + \frac{a_3^2}{a_3 + b_3}$$

Introducing the new positive variables $p = a_1 + a_2$, $q = b_1 + b_2$ and using the already proved case for $n = 2$, results in

$$\frac{p^2}{p + q} + \frac{a_3^2}{a_3 + b_3} \geq \frac{\left(p + a_3\right)^2}{p + q + a_3 + b_3} = \frac{\left(a_1 + a_2 + a_3\right)^2}{a_1 + b_1 + a_2 + b_2 + a_3 + b_3} \qquad (10.4)$$

As a result,

$$\frac{a_1^2}{a_1 + b_1} + \frac{a_2^2}{a_2 + b_2} + \frac{a_3^2}{a_3 + b_3} \geq \frac{\left(a_1 + a_2 + a_3\right)^2}{a_1 + b_1 + a_2 + b_2 + a_3 + b_3} \qquad (10.5)$$

so the case $n = 3$ has also been proved. Continuing this reasoning proves inequality (10.1) inductively, for any $n > 3$.

The left and right side of inequality (10.1), multiplied by an appropriate factor, can be physically interpreted as kinetic energy after a perfectly inelastic collision.

Indeed, if an object with mass a, moving horizontally (parallel to the ground) with velocity v_0, collides with a stationary object with mass b and the collision is perfectly inelastic, a single object with mass $a + b$ is formed after the collision, moving with velocity v (Figure 10.1a).

According to the law of conservation of the linear momentum, the sum of the two momenta before collision is equal to their sum after collision:

$$av_0 + 0 = \left(a + b\right)v$$

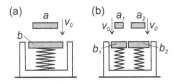

FIGURE 10.1 (a) A perfect inelastic collision between an object with mass a, moving with constant velocity v_0, and a stationary object with mass b. (b) A perfect inelastic collision between a pair of objects with masses a_1 and a_2 moving with constant velocity v_0 and a pair of stationary objects with masses b_1 and b_2.

from which, the velocity v of the two objects after the inelastic collision is given by

$$v = \frac{av_0}{a+b} \tag{10.6}$$

The kinetic energy of the system after the inelastic impact is therefore equal to

$$E_k = \frac{1}{2}(a+b)v^2 = \frac{a^2 v_0^2}{2(a+b)}$$

Therefore, the right-hand side of inequality (10.1) multiplied by the constant $v_0^2 / 2$ gives the kinetic energy after the inelastic impact of objects with masses a and b where v_0 is velocity of the moving object.

Suppose that the object with mass a has been segmented into two objects with masses a_1 and a_2 ($a = a_1 + a_2$) and the second object with mass b has also been segmented into two objects with masses b_1 and b_2 ($b = b_1 + b_2$) (Figure 10.1). Multiplying both sides of inequality (10.2) by the factor $v_0^2 / 2$ yields

$$\frac{a_1^2 v_0^2}{2(a_1 + b_1)} + \frac{a_2^2 v_0^2}{2(a_2 + b_2)} \geq \frac{(a_1 + a_2)^2 v_0^2}{2(a_1 + b_1 + a_2 + b_2)} \tag{10.7}$$

and makes the resultant inequality (10.7) reverse engineerable. Both sides of inequality (10.7) effectively describe the same output (total kinetics energy) associated with the design options in Figure 10.1a and b.

The left-hand side of inequality (10.7) is the total kinetic energy after a perfectly inelastic collision of two objects with masses a_1 and a_2 moving with velocity v_0 towards two stationary objects with masses b_1 and b_2 (Figure 10.1b). The right-hand side of inequality (10.7) is the kinetic energy after a perfectly inelastic collision of a single object with mass equal to the combined mass of the two objects, moving with velocity v_0 towards a single object with mass equal to the combined mass of the stationary objects (Figure 10.1a).

The reverse engineering of inequality (10.7) predicts that aggregating objects colliding perfectly inelastically results in a smaller total kinetic energy after the inelastic collision (Todinov, 2022d).

Now consider the total kinetic energy $(1/2)a_1 v_0^2 + (1/2)a_2 v_0^2$ of the two moving objects before the inelastic collision. It is equal to the kinetic energy $(1/2)av_0^2$ before the collision of the single moving object:

$$(1/2)a_1 v_0^2 + (1/2)a_2 v_0^2 = (1/2)av_0^2 \tag{10.8}$$

Multiplying inequality (10.7) by '−1' and adding to both sides Equation (10.8) results in the inequality

$$\left(\frac{a_1 v_0^2}{2} - \frac{a_1^2 v_0^2}{2(a_1 + b_1)} \right) + \left(\frac{a_2 v_0^2}{2} - \frac{a_2^2 v_0^2}{2(a_2 + b_2)} \right)$$

$$\leq \frac{(a_1 + a_2) v_0^2}{2} - \frac{(a_1 + a_2)^2 v_0^2}{2(a_1 + b_1 + a_2 + b_2)} \tag{10.9}$$

Both sides of inequality (10.9) represent the absorbed kinetic energy during inelastic collision, characterising the design options in Figure 10.1a and b. The left-hand side of inequality (10.9) is the absorbed kinetic energy during the inelastic collision of the two pairs of objects, while the right-hand side is the absorbed kinetic energy during the inelastic collision of the single objects.

Inequality (10.9) yields the prediction that a perfectly inelastic impact between single objects is associated with a greater amount of absorbed kinetic energy compared to the perfectly inelastic collision between the parts of the segmented objects. Aggregating objects, therefore, increases the absorbed kinetic energy during inelastic collision (Todinov, 2022d). The prediction obtained from inequality (10.9) can be applied as a basis of a strategy for mitigating shocks from inelastic impacts.

It may seem that the systems in Figure 10.1 are two different systems. In fact, Figures 10.1a and b are alternative designs of the same shock-absorbing system. Aggregating the impacting masses increases the absorbed kinetic energy during inelastic collision and the design option in Figure 10.1a should be preferred.

It needs to be pointed out that aggregation does not always achieve an increase of the absorbed kinetic energy for the system in Figure 10.1. Thus if $a_1/b_1 = a_2/b_2$, the right-hand side of Equation (10.3) is zero, equality is attained in inequality (10.2) and no increase in the absorbed kinetic energy is present after the inelastic collision.

Asymmetry must be present for an increase of the absorbed kinetic energy to occur. Again, the requirement for asymmetry to increase the absorbed kinetics energy is rather counterintuitive and makes this prediction difficult to make by alternative means bypassing inequality (10.1).

10.2 RANKING THE STIFFNESS OF ALTERNATIVE MECHANICAL ASSEMBLIES THROUGH REVERSE ENGINEERING OF AN ALGEBRAIC INEQUALITY

A special case of the general super-additive inequality (2.39) is the algebraic inequality

$$\frac{a_1 b_1}{a_1 + b_1} + \frac{a_2 b_2}{a_2 + b_2} + \ldots + \frac{a_n b_n}{a_n + b_n} \leq \frac{(a_1 + a_2 + \ldots + a_n)(b_1 + b_2 + \ldots + b_n)}{(a_1 + a_2 + \ldots + a_n) + (b_1 + b_2 + \ldots + b_n)} \tag{10.10}$$

where both controlling factors $a = a_1 + a_2 + \ldots + a_n$ and $b = b_1 + b_2 + \ldots + b_n$ are additive positive quantities.

This inequality can be proved by a direct manipulation which reduces it to a standard inequality already considered in Chapter 2. The left-hand side of inequality (10.10) can be presented as

$$\sum_{i=1}^{n} \frac{a_i b_i}{a_i + b_i} = \frac{a_1 b_1}{a_1 + b_1} - a_1 + \frac{a_2 b_2}{a_2 + b_2} - a_2 + \ldots + \frac{a_n b_n}{a_n + b_n} - a_n + \left(a_1 + \ldots + a_n\right) \quad (10.11)$$

Conducting the subtractions $\dfrac{a_i b_i}{a_i + b_i} - a_i$ transforms (10.11) into the equality

$$\sum_{i=1}^{n} \frac{a_i b_i}{a_i + b_i} = \left(a_1 + \ldots + a_n\right) - \frac{a_1^2}{a_1 + b_1} - \frac{a_2^2}{a_2 + b_2} - \ldots - \frac{a_n^2}{a_n + b_n} \quad (10.12)$$

According to the Bergström inequality,

$$\frac{a_1^2}{a_1 + b_1} + \frac{a_2^2}{a_2 + b_2} + \ldots + \frac{a_n^2}{a_n + b_n} \geq \frac{\left(a_1 + \ldots + a_n\right) \times \left(a_1 + \ldots + a_n\right)}{\left(a_1 + \ldots + a_n\right) + \left(b_1 + \ldots + b_n\right)}$$

Consequently, after the substitution of the right-hand side of this inequality in the right-hand side of equality (10.12), the next inequality is obtained:

$$\sum_{i=1}^{n} \frac{a_i b_i}{a_i + b_i} \leq \left(a_1 + \ldots + a_n\right) - \frac{\left(a_1 + \ldots + a_n\right) \times \left(a_1 + \ldots + a_n\right)}{\left(a_1 + \ldots + a_n\right) + \left(b_1 + \ldots + b_n\right)}$$

$$= \frac{\left(a_1 + \ldots + a_n\right) \times \left(b_1 + \ldots + b_n\right)}{\left(a_1 + \ldots + a_n\right) + \left(b_1 + \ldots + b_n\right)}$$

which completes the proof of inequality (10.10).

Inequality (10.10) has a natural reverse engineering. Suppose that there are n pairs of elastic elements A_i, B_i, and a_i stands for the stiffness of elastic element A_i while b_i stands for the stiffness of elastic element B_i, $i = 1, \ldots, n$. Noticing that, in this case, $\dfrac{a_i b_i}{a_i + b_i}$ is the equivalent stiffness of the elastic elements A_i and B_i connected in series, inequality (10.10) yields an interesting prediction. The equivalent stiffness of n pairs of elastic elements A_i, B_i working in parallel, where the elastic elements A_i, B_i in each pair are connected in series (Figure 10.2a), does not exceed the equivalent stiffness of the system in Figure 10.2b where all A_i and B_i elements are connected in parallel and the two assemblies are connected in series (Todinov, 2022d).

However, for $b_i/a_i = k$, $i = 1, 2, \ldots, n$, it can be shown that the two assemblies have the same stiffness and are equivalent.

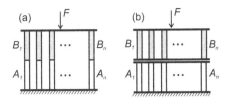

FIGURE 10.2 Two alternative assemblies with different arrangement of the elastic elements: (a) without bracing plates in the middle, (b) with bracing plates in the middle connecting elastic elements working in parallel.

Indeed, for $b_i = ka_i$, the substitution in the left-hand side of inequality (10.10) yields:

$$\frac{a_1 k}{1+k} + \frac{a_2 k}{1+k} + \ldots + \frac{a_n k}{1+k} = \frac{\left(a_1 + a_2 + \ldots + a_n\right)k}{1+k} \tag{10.13}$$

The substitution in the right-hand side of inequality (10.10) results in

$$\frac{\left(a_1 + a_2 + \ldots + a_n\right)\left(b_1 + b_2 + \ldots + b_n\right)}{\left(a_1 + a_2 + \ldots + a_n\right) + \left(b_1 + b_2 + \ldots + b_n\right)}$$

$$= \frac{\left(a_1 + a_2 + \ldots + a_n\right)^2 k}{\left(a_1 + a_2 + \ldots + a_n\right) + k\left(a_1 + a_2 + \ldots + a_n\right)} = \frac{\left(a_1 + a_2 + \ldots + a_n\right)k}{1+k} \tag{10.14}$$

The right-hand sides of equalities (10.13) and (10.14) are identical; therefore, the two assemblies are equivalent (have the same stiffness).

The important conclusion from the reverse engineering of inequality (10.10) is that the increase of stiffness associated with the assembly in Figure 10.2b *is lost completely if the pairs are characterised by the same ratio of the stiffness values of the individual elastic elements.*

This is a counter-intuitive statement which can be demonstrated on two pairs of elastic elements (A_1, B_1), (A_2, B_2) with the same length in unloaded state and stiffness values: $a_1 = 1400 N/m$, $b_1 = 800 N/m$, $a_2 = 1050 N/m$, $b_2 = 600 N/m$, respectively (Figure 10.3a and b).

Because $a_1/b_1 = a_2/b_2 = 1.75$,

$$\frac{a_1 b_1}{a_1 + b_1} + \frac{a_2 b_2}{a_2 + b_2} = \frac{\left(a_1 + a_2\right)\left(b_1 + b_2\right)}{\left(a_1 + a_2\right) + \left(b_1 + b_2\right)} = 890.91 N/m$$

and the assemblies in Figures 10.3a and 10.3b have the same stiffness. However, for the pairs of elastic elements $a_1 = 1400 N/m$, $b_1 = 800 N/m$ and $a_2 = 600 N/m$,

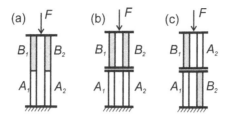

FIGURE 10.3 Alternative assemblies with different arrangement of the elastic components: (a) two pairs of elastic components connected in parallel in which two elastic components of types A and B are connected in series; (b) two elastic assemblies connected in series; one of the assemblies includes two elastic components of type A connected in parallel and the other assembly includes two elastic components of type B connected in parallel; (c) two elastic assemblies connected in series. Each of the assemblies includes elastic components of type A and type B connected in parallel.

$b_2 = 1050 N/m$ (Figure 10.3c) the stiffness values in the pairs are not proportional: $a_1/b_1 = 1.75 \neq a_2/b_2 = 0.57$. As a result, the left- and right-hand side of inequality (10.10) are no longer equal:

$$\frac{a_1 b_1}{a_1 + b_1} + \frac{a_2 b_2}{a_2 + b_2} = 890.91\, N/m < \frac{(a_1 + a_2)(b_1 + b_2)}{(a_1 + a_2) + (b_1 + b_2)} = 961\, N/m$$

and the result is an increase of the stiffness for the assembly in Figure 10.3c compared to the assemblies in Figure 10.3a and b. Inequality (10.10) can be used as a basis of a strategy for increasing the stiffness of mechanical assemblies similar to the ones in Figure 10.2a and b.

10.3 REVERSE ENGINEERING OF SINGLE-VARIABLE SUPER-ADDITIVE AND SUB-ADDITIVE INEQUALITIES

10.3.1 Single-Variable Sub-Additive and Super-Additive Inequalities

Consider an additive quantity x that has been segmented into a number of non-negative segments $x_1, x_2, \ldots, x_n$ ($x = x_1 + x_2 + \ldots + x_n$). The output associated with each of the segments x_i is given by $f(x_i)$. According to Section 2.2.3, if the function $f(x)$ is a *sub-additive* function of the additive quantity x, for the parts x_1, $x_2, \ldots, x_n$ into which the additive quantity x has been segmented, the next inequality holds:

$$f(x_1 + x_2 + \ldots + x_n) \leq f(x_1) + f(x_2) + \ldots + f(x_n) \tag{10.15}$$

for any set of non-negative segments $x_1, x_2, \ldots, x_n$.

According to Section 2.2.3, if the function $f(x)$ is a *super-additive* function of the additive quantity x, for the non-negative parts $x_1, x_2, ..., x_n$ into which the additive quantity x has been segmented, the next inequality holds:

$$f\left(x_1 + x_2 + ... + x_n\right) \geq f\left(x_1\right) + f\left(x_2\right) + ... + f\left(x_n\right) \qquad (10.16)$$

for any set of non-negative $x_1, x_2, ..., x_n$.

Inequalities (10.15) and (10.16) have important potential applications in optimising processes. Suppose that the function $f(x)$ measures the effect/output of a particular additive quantity, the variables x_i ($i = 1, ..., n$) denote the different segments of the quantity x, and the effects $f(x_i)$ are also additive quantities. The reverse engineering of inequalities (10.15) and (10.16) provides the unique opportunity to increase the effect of the additive quantity by segmenting or aggregating it, depending on whether the function $f(x)$ is concave or convex. If the function is concave, with a domain $[0, \infty)$ and range $[0, \infty)$, the function is sub-additive and segmenting the additive quantity x results in a larger output. If the function $f(x)$ is convex, with a domain $[0, \infty)$ and if $f(0) \leq 0$, the function is super-additive and aggregating the additive quantity results in a larger output (see Chapter 2, Section 2.2.3). Inequalities (10.15) and (10.16) have universal application in science and technology as long as, x_i and the terms $f(x_i)$ are additive quantities and have a meaningful physical interpretation.

The use of inequalities based on single-variable sub- and super-additive functions in process optimisation will be illustrated by power-type dependences involving a single controlling factor. Power functions are widespread (Andriani and McKelvey, 2007; Newman, 2007; Easley and Kleinberg, 2010). Many additive quantities can be approximated very well by power functions of the type

$$y = ax^p \qquad (10.17)$$

where a and p are constants ($a \neq 0$; $p > 0$), x is the controlling factor ($x \geq 0$) and y is the output quantity.

The power functions given by Equation (10.17) are encountered frequently in mathematical modelling, where y stands for frequency, energy, power, force, damage, profit, pollution, etc. Power laws appear, for example, in cases where positive feedback loops are determining the output. Often, at the heart of these positive feedback loops is the *preferential attachment phenomenon*, according to which a commodity is distributed according to how much commodity is already present.

Depending on whether the power p in (10.17) is greater or smaller than unity, sub- or super-additive inequalities can be used to perform segmentation or aggregation of the additive controlling factor in order to attain enhanced performance.

The additive output quantity y can be a convex or concave function of the controlling factor x ($x \geq 0$) depending on whether the second derivative $d^2y/dx^2 = ap$ $(p - 1)x^{p-2}$ with respect to x is positive or negative. This in turn depends on whether the power p is greater or smaller than 1. If $p > 1$, $d^2y/dx^2 \geq 0$, and the power function y is convex. If $0 < p < 1$, $d^2y/dx^2 \leq 0$ and the power function y is concave. If $p < 0$, $d^2y/dx^2 = ap(p-1)x^{p-2} > 0$ and the power function y is convex.

Suppose that the output function y is of the type presented by Equation (10.17). For a controlling factor x varying in the interval $[0, \infty)$ the function (5.17) is strictly convex if $p > 1$ and the following super-additive inequality holds:

$$a\left(x_1 + x_2 + \ldots + x_n\right)^p > ax_1^p + ax_2^p + \ldots + ax_n^p \qquad (10.18)$$

This inequality can be proved easily by a mathematical induction by proving first the base case corresponding to $n = 2$ (details of the proof have been omitted). The physical interpretation of the inequality states that aggregating the non-zero additive factors x_i yields a larger total effect.

If $0 < p < 1$, the power function (10.17) is strictly concave and the following sub-additive inequality holds:

$$a\left(x_1 + x_2 + \ldots + x_n\right)^p < ax_1^p + ax_2^p + \ldots + ax_n^p \qquad (10.19)$$

which means that segmenting the additive factor x into several non-zero segments x_i yields a larger total effect.

The algebraic inequality (10.19) always holds if

$$p < 1.$$

Proof: Inequality (10.19) can be rigorously proved by induction. For $n = 2$, and $x_1 > 0$, $x_2 > 0$, the inequality reduces to

$$a\left(x_1 + x_2\right)^p < ax_1^p + ax_2^p \qquad (10.20)$$

We will use an argument based on the fact that $f(x) = ax^p$ is a strictly concave function. Note that $w_1 = x_1/(x_1 + x_2)$ and $w_2 = x_2/(x_1 + x_2)$ can be treated as weights because $0 \leq w_1, w_2 \leq 1$ and $w_1 + w_2 = 1$. Because $f(x) = ax^p$ is a strictly concave function, for the values $x = x_1 + x_2$ and $x = 0$, $(x_1 + x_2 <> 0)$ the Jensen's inequality for concave functions (see Section 2.1.3) gives:

$$f\left(x_1\right) = f\left(\frac{x_1}{x_1 + x_2} \times \left(x_1 + x_2\right) + \frac{x_2}{x_1 + x_1} \times 0\right) > \frac{x_1}{x_1 + x_2} f\left(x_1 + x_2\right)$$

$$+ \frac{x_2}{x_1 + x_2} f(0) \qquad (10.21)$$

$$f\left(x_2\right) = f\left(\frac{x_2}{x_1 + x_2} \times \left(x_1 + x_2\right) + \frac{x_1}{x_1 + x_2} \times 0\right) > \frac{x_2}{x_1 + x_2} f\left(x_1 + x_2\right)$$

$$+ \frac{x_1}{x_1 + x_2} f(0) \qquad (10.22)$$

Adding inequalities (10.21) and (10.22) yields

$$f(x_1) + f(x_2) > f(x_1 + x_2) + f(0) \qquad (10.23)$$

Since $f(0) = a \times 0^p = 0$ it follows that $f(x_1) + f(x_2) > f(x_1 + x_2)$, which completes the proof of inequality (10.20).

Now suppose that inequality (10.19) holds for $n = k$, where $x_i > 0$ (induction hypothesis):

$$a(x_1 + x_2 + ... + x_k)^p < ax_1^p + ax_2^p + ... + ax_k^p \qquad (10.24)$$

Setting $x' = x_1 + x_2 + ... + x_k$, according to inequality (10.20), the inequality

$$a(x' + x_{k+1})^p < a(x')^p + ax_{k+1}^p \qquad (10.25)$$

holds true, where $x_{k+1} > 0$.

Considering that according to the induction hypothesis: $a(x')^p < ax_1^p + ax_2^p + ... + ax_k^p$, inequality (10.25) transforms into

$$a(x_1 + x_2 + ... + x_k + x_{k+1})^p < ax_1^p + ax_2^p + ... + ax_k^p + ax_{k+1}^p \qquad (10.26)$$

which proves the induction step. This, together with the base case corresponding to $n = 2$, proves inequality (10.19) for any $n \geq 2$.

Inequality (10.18) can be proved by using similar reasoning and details will be omitted.

10.3.2 REVERSE ENGINEERING OF A SUB-ADDITIVE ALGEBRAIC INEQUALITY TO IMPROVE THE PERFORMANCE OF A FILTER

Asymmetric output characteristics can be used to improve the performance of filters thereby improving their collection efficiency and reducing the risk of pollution (Todinov, 2024b).

According to Fu et al. (2010), the dependence of the collection efficiency and thickness of the fibrous media in the filter, for a specified duration, is described by a nonlinear, concave curve reaching a region of saturation with increasing the thickness of the filter (Figure 10.4).

A plot collection efficiency [%] versus log-thickness based on experimental results published in Fu et al. (2010) can be fitted very well with a straight line which shows that in the range 8–300 μm, the trend "Collection efficiency [%] – Thickness of filter in μm" can be approximated very well by a concave power law with power approximately equal to $p \approx 0.27$.

The concave function associated with the concave collection efficiency characteristics in Figure 10.4 can then be exploited to increase the performance of such filters through reverse engineering of an algebraic inequality.

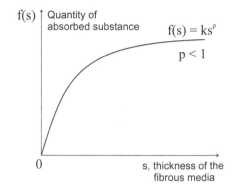

FIGURE 10.4 Quantity of absorbed substance (collection efficiency) versus thickness of the filter.

Let the quantity $f(s)$ describe the collection efficiency of the filter and s stand for the thickness of the filter. Let the concave curve $f(s)$ be approximated by the concave power law

$$f(s) = ks^p \tag{10.27}$$

where $p < 1$ is the exponent of the power law and $k > 0$ is a constant.

According to inequality (10.19), the quantity of absorbed harmful substance can be increased by segmenting the filter (Figure 10.5). This is because of the asymmetric concave dependence (10.27).

Segmenting the filter with thickness s into smaller filters with thicknesses $s_1, s_2, \ldots,$ s_n ($s_i > 0$) with the same total thickness of the fibrous media (Figure 10.4) ($s_1 + s_2 + \ldots + s_n = s$) yields a greater collection efficiency and reduces the risk of pollution. The quantities f_i of absorbed substance by the segmented filters are given by

$$f_1 = ks_1^p, f_2 = ks_2^p, \ldots, f_n = ks_n^p \tag{10.28}$$

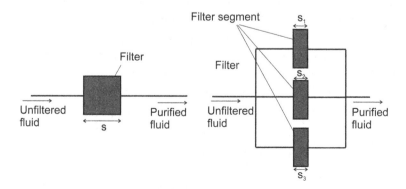

FIGURE 10.5 (a) Original and (b) segmented filter with greater collection efficiency.

and their sum results in a larger total quantity of absorbed substance because of the algebraic inequality:

$$k\left(s_1 + s_2 + \ldots + s_n\right)^p < ks_1^p + ks_2^p + \ldots + ks_n^p \tag{10.29}$$

If the collection efficiency of the filter were symmetric (linear), there would not be any benefits for the collection efficiency. Indeed for $p = 1$, inequality (10.29) transforms into an equality:

$$k\left(s_1 + s_2 + \ldots + s_n\right) = ks_1 + ks_2 + \ldots + ks_n \tag{10.30}$$

Similarly, if the collection efficiency of the filter were non-linear and convex (bending upwards), there would be no benefits for the collection efficiency. In fact, there would be a reduction in the collection efficiency due to segmentation. Indeed, in this case, $p > 1$, and

$$k\left(s_1 + s_2 + \ldots + s_n\right)^p > ks_1^p + ks_2^p + \ldots + ks_n^p \tag{10.31}$$

10.3.3   REVERSE ENGINEERING OF A SUPER-ADDITIVE INEQUALITY TO MINIMISE THE FORMATION OF BRITTLE PHASE DURING SOLIDIFICATION

Consider an application example (Todinov, 2020c) involving the formation of undesirable brittle phase, during the solidification of a specimen with spherical shape and volume V. The specimen solidifies from a particular initial temperature of the molten alloy. The quantity z of undesirable brittle phase formed during solidification (which is an additive quantity) is given by the power function:

$$z = aV^p \tag{10.32}$$

where a and $p > 1$ are constants which depend on the temperature, the nature of the alloy and the shape of the specimen. The controlling factor in the power-law dependence (10.32) is the volume V of the specimen which is also an additive quantity. The formation of brittle phase during solidification compromises the mechanical properties of the specimen and an increased quantity of brittle phase means significantly reduced strength.

The additive quantity z of brittle phase is a convex function of the volume of the specimen because the second derivative with respect to the volume V is positive: $d^2z/dV^2 = ap(p-1)V^{p-2}$. Since function (10.32) is strictly convex, the quantity z varies in the interval $[0, \infty)$ and $z(0) = 0$, the function (10.32) is super-additive. According to inequality (10.18), the inequality

$$a\left(V_1 + V_2 + \ldots + V_n\right)^p > aV_1^p + aV_2^p + \ldots + aV_n^p \tag{10.33}$$

holds true. The right side of inequality (10.33) can be interpreted as the sum of the quantities of brittle phase in n specimens with non-zero volumes $V_1, V_2, ..., V_n$. The left-hand side of inequality (10.33) can be interpreted as quantity of brittle phase in a single specimen with a similar shape and volume equal to the sum of the volumes $V_1, V_2, ..., V_n$ of the smaller specimens.

As a result, inequality (10.33) predicts that segmenting a specimen with volume V into n specimens with a similar shape and volumes $V_1, V_2, ..., V_n$, ($V = V_1 + V_2 + ... + V_n$) will decrease the amount of unwanted brittle phase during solidification, irrespective of the individual volumes $V_1, V_2, ..., V_n$.

10.3.4 REVERSE ENGINEERING OF A SUPER-ADDITIVE INEQUALITY TO MINIMISE THE DRAG FORCE EXPERIENCED BY AN OBJECT MOVING THROUGH FLUID

This example illustrates generating knowledge through reverse engineering of super-additive inequalities and using it to reduce the drag force F_d acting on a body with volume V moving through fluid with a constant velocity v (Todinov, 2020c). The drag force is an additive property specified by

$$F_d = (1/2)C_d \rho v^2 A \qquad (10.34)$$

where C_d is the drag coefficient—a dimensionless number that depends on the shape of the body; ρ [kg/m³] is the density of the fluid, v [m/s] is the velocity of the body and A [m²] is the cross-sectional area of the body. Considering that for a body with volume V, the cross-sectional area is proportional to $V^{2/3}$, the drag force can be presented as a function of the additive property 'volume V' of the body:

$$F_d = aV^{2/3} \qquad (10.35)$$

where a is a constant.

The drag force given by Equation (10.35) is a strictly concave function because the second derivative with respect to the volume V is negative: $d^2F_d/dV^2 = -2a/(9V^{4/3})$. Because for volume varying in the interval $[0, \infty)$ the drag force varies in the range $[0, \infty)$, the drag force function (10.35) is sub-additive. According to inequality (10.19), the inequality

$$a(V_1 + V_2 + ... + V_n)^{2/3} < aV_1^{2/3} + aV_2^{2/3} + ... + aV_n^{2/3} \qquad (10.36)$$

holds true. The right-hand side of inequality (10.36) is an additive quantity and can be physically interpreted as the total drag force acting on n bodies with non-zero volumes $V_1, V_2, ..., V_n$ moving at the same speed. The left-hand side of inequality (10.36) is also an additive quantity and can be interpreted as the drag force acting on a body with a similar shape and volume equal to the sum of the volumes $V_1, V_2, ..., V_n$ of the separate bodies and moving at the same speed.

As a result, inequality (10.36) predicts that aggregating n bodies with volumes $V_1, V_2, ..., V_n$ into a single large body with a similar shape and larger volume $V = V_1 + V_2 + ... + V_n$, equal to the sum of the volumes of the n bodies, decreases the total drag force, irrespective of the actual volumes $V_1, V_2, ..., V_n$.

10.3.5   Reverse Engineering of a Sub-Additive Inequality to Maximise the Profit from an Investment

Consider now an application example from economics involving a sub-additive inequality. Suppose that the annual profit z from an investment in a particular enterprise is given by the power function:

$$z = c x^q \qquad (10.37)$$

where x is the size of the investment, c and $q < 1$ are constants which depend on the particular enterprise. The additive factor in the power-law dependence (10.37) is the size of the investment x. The profit z is also an additive quantity.

For $0 < q < 1$, the profit z is a concave function of the size of investment because the second derivative with respect to the investment x is negative: $d^2z/dx^2 = cq(q - 1)x^{q-2} < 0$. Because for investment varying in the interval $[0, \infty)$, the profit varies in the range $[0, \infty)$, according to inequality (10.19), the following inequality holds:

$$c\left(x_1 + x_2 + ... + x_n\right)^q < cx_1^q + cx_2^q + ... + cx_n^q \qquad (10.38)$$

Inequality (10.38) predicts that splitting the initial investment x and investing in n parallel enterprises of the same type will result in larger profit than investing the entire sum x in a single enterprise. The difference in profit can be significant as is shown in the next numerical example. Thus, for a profit dependence

$$f\left(x\right) = 15.3x^{0.42} \qquad (10.39)$$

the profit from investing $x = \$10000$ is $15.3 \times 10000^{0.42} = \$732.3$. Splitting the investment in two and investing \$5000 in two parallel enterprises yield profit of magnitude $15.3 \times 5000^{0.42} + 15.3 \times 5000^{0.42} = \$1094.7$ which is 1.49 times larger than the profit obtained from the single investment.

11 Enhancing Decision-Making by Reverse Engineering of Algebraic Inequalities

11.1 REVERSE ENGINEERING OF AN ALGEBRAIC INEQUALITY RELATED TO RANKING THE MAGNITUDES OF SEQUENTIAL RANDOM EVENTS

Algebraic inequalities do not need to be complex, in order to extract new knowledge from their physical interpretation. The next example, involving the reverse engineering of a very simple algebraic inequality with far reaching consequences, has been prompted by the short note by Cover (1987). Consider the simple algebraic inequality (Todinov, 2021a)

$$p^2 + (1-p)^2 + 4p(1-p) \geq 1 \tag{11.1}$$

where $0 \leq p \leq 1$. This inequality can be proved easily by presenting the left-hand side as

$$p^2 + (1-p)^2 + 4p(1-p) = \left[p + (1-p)\right]^2 + 2p(1-p) = 1 + 2p(1-p)$$

The left-hand side of inequality (11.1), equal to $1 + 2p(1-p)$, is obviously not smaller than one because $p(1-p) \geq 0$. If both sides of inequality (11.1) are divided by 2, inequality (11.1) is transformed into the next reverse engineerable equivalent inequality:

$$(1/2)p^2 + (1/2)(1-p)(1-p) + (1-p)p + p(1-p) \geq 1/2 \tag{11.2}$$

Inequality (11.2) can be interpreted by creating physical meaning for the variable p and for the terms in the left-hand side of the inequality. Because $0 \leq p \leq 1$, it is natural to interpret the variable p as 'probability'. If p is the probability that the magnitude y of a random event with a uniform distribution exceeds a specified threshold C (Figure 11.1), the terms p^2, $(1-p)^2$, $(1-p)p$, $p(1-p)$ can be easily interpreted.

The term p^2 represents the probability that the magnitudes of two statistically independent random events, each following the same uniform distribution, will

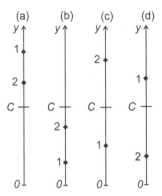

FIGURE 11.1 Possible locations of the magnitudes of two random events with respect to a threshold C: (a) both magnitudes above the pre-selected threshold C; (b) both magnitudes below the pre-selected threshold C; first event magnitude below the threshold C, second event magnitude above the threshold C; (d) first event magnitude above the threshold C, second event magnitude below the threshold C.

both exceed the fixed threshold C (Figure 11.1a). The term $(1 - p)^2$ represents the probability that the magnitudes of both random events will be below the threshold C (Figure 11.1b). The term $(1 - p)p$ represents the probability that the magnitude of the first random event will be below the threshold C while the magnitude of the second random event will be above the threshold C (Figure 11.1c). Finally, $p(1 - p)$ represents the probability that the magnitude of the first random event will be above the threshold C and the magnitude of the second random event will be below the threshold C (Figure 11.1d).

The first term $(1/2)p^2$ in inequality (11.2) can then be interpreted as the probability of a compound event where the magnitudes of two sequential random events both exceed the selected threshold C and the magnitude of the second random event is smaller than that of the first (Figure 11.1a).

The second term $\frac{1}{2}(1 - p)(1 - p)$ in inequality (11.2) can be interpreted as the probability of a compound event where the magnitudes of two sequential random events are both below the selected threshold C and the magnitude of the second random event is larger than that of the first (Figure 11.1b).

The third term $(1 - p)p$ in inequality (11.2) can be interpreted as the probability of a compound event where the magnitude of the first random event is below the threshold C while the magnitude of the second random event is above the threshold C (Figure 11.1c). Clearly, in this case, the magnitude of the second event is certainly larger than that of the first event.

Finally, the fourth term $p(1 - p)$ in inequality (11.2) can be interpreted as the probability of a compound event where the magnitude of the first random event is above the threshold C while the magnitude of the second random event is below the threshold C (Figure 11.1d). Clearly, in this case, the magnitude of the first event is certainly larger than that of the second event.

Consider a real-world problem related to predicting the magnitude ranking of safety-critical random events, which is crucial for determining whether additional resources are needed to mitigate the consequences of future random events. Examples of such random events include crude oil leaks into marine water from subsea oil and gas production pipelines, the magnitudes of floods in consecutive years and the severity of fires. If no information is available about the distribution of the event magnitudes, the probability of correctly predicting whether the magnitude of the second event will exceed that of the first appears to be only 50%, with no apparent way to improve this likelihood.

According to a prediction strategy outlined in (Cover, 1987), if the magnitude of the first random event is larger than the pre-selected comparison threshold C (Figures 11.1a and 11.1d), a prediction is made that the second random event will have a smaller magnitude than the magnitude of the first random event. If the magnitude of the first random event is smaller than the comparison threshold C (Figures 11.1b and 11.1c), a prediction is made that the second random event will have a larger magnitude than the magnitude of the first event.

According to inequality (11.2), by following this prediction strategy, the probability of predicting correctly the magnitude ranking of the random events can be made to be greater than 50%.

Indeed, when both random events have magnitudes either larger or smaller than the pre-selected threshold C, the likelihood of correctly predicting the magnitude of the second event is 0.5. If the first random event has a magnitude smaller than the threshold C and the second event has a magnitude larger than C, the likelihood of a correct prediction is 1.0. Similarly, if the first event has a magnitude larger than the threshold C and the second event has a magnitude smaller than C, the likelihood of a correct prediction is also 1.0.

Contrary to the statement made in (Cover, 1987) however, from inequality (11.2), it follows that the prediction strategy described earlier *does not ensure correct prediction with probability strictly greater than 0.5*.

If the threshold C were chosen to be too large or too small within the physically possible range of values for the magnitude of the random event, it would result in underestimating or overestimating the expected event magnitude and $1 - p \approx 0$ or $p \approx 0$. Consequently, the quantity $p(1 - p)$ on the left-hand side of inequality (11.2) would be very small, making it practically impossible to achieve a probability of correct prediction greater than 0.5.

The maximum product of two positive values with a given sum occurs when the values are equal. Consequently, the product pq where $p + q = 1$, attains maximum when $p = q = 1/2$. Consequently, the maximum possible value of $p(1 - p)$ is 0.25 which cannot be exceeded. As a result, the optimal threshold C is positioned such that the probability that the event magnitude is less than C equals the probability that it is greater than C. Under these conditions, theoretically correct predictions will occur in 75% of trials. In reality, the threshold position of equal probability is not known and the threshold is selected randomly within the physically possible range of values for the magnitude of random events.

To evaluate the described decision strategy, a Monte-Carlo simulation experiment was conducted (Todinov, 2021a). Consider a scenario where crude oil leaks from a pipeline with a maximum flow rate of 300 l/s. Given that leaks cannot be negative and cannot exceed this maximum, the plausible range for all potential leaks spans from 0 to 300 l/s. The specific magnitudes within this physically feasible range are uncertain, leading to the assumption of a uniform distribution for leak magnitudes. As a result, the lower and upper limits are selected randomly within the permissible physical range [0, 300].

A series of Monte-Carlo simulations, each comprising 10 million random trials, yielded probabilities of correct predictions ranging from 0.50 to 0.64. These results confirm that the proposed strategy effectively reduces uncertainty and provides predictions regarding the magnitude ranking of subsequent leaks that surpass random guessing. Without a preliminary threshold comparison, random predictions would be characterised by an exact 50% probability of being correct. Moreover, by reinterpreting the random events, the abstract inequality (11.2) can be reverse engineered to cover other real-world processes such as magnitudes of random shocks affecting devices or structures. In some cases, this kind of reverse engineering could also help in transforming a neutral bet—expected to yield zero profit—into a profitable bet with positive expected return. Accumulating numerous such bets leads to a positive net gain (Todinov, 2013).

11.2 IMPROVING PRODUCT RELIABILITY BY INCREASING THE LEVEL OF BALANCING

Well-balanced systems and assemblies distribute the load more uniformly across components and exhibit higher reliability. Improving the level of balancing in systems and assemblies can be achieved by

- Ensuring more uniform load distribution among components
- Ensuring conditions for self-balancing
- Reducing the variability of risk-critical parameters

11.2.1 ENSURING MORE UNIFORM LOAD DISTRIBUTION AMONG COMPONENTS

Ensuring a more uniform distribution of the load is essential for both electrical and mechanical components. For instance, in mechanical assemblies, the use of splines instead of keys for transmitting torque from a rotating shaft to a gear, sprocket, or pulley, significantly improves load distribution. This not only reduces stress but also strengthens the connection and improves its resistance to failure under overload conditions.

Similarly, consider the impact of fasteners in a flange connection: a small number of fasteners can lead to excessive stress. Increasing the number of fasteners not only balances the load distribution but also lowers the risk of failure.

Segmenting load-carrying components often increases contact area, ensuring better conformity with other components. This reduces contact stresses and enhances reliability.

Improved load distribution through segmentation also enhances balance in rotating components, resulting in greater stability, reduced vibration amplitudes, smaller inertia forces and improved reliability. This principle applies to designs such as engines with multiple cylinders or turbines with multiple blades.

11.2.2 ENSURING CONDITIONS FOR SELF-BALANCING

Unbalanced forces contribute to premature wear, fatigue degradation and system failure. Typically, enhancing the balance level within a system reduces force magnitudes, loading stresses and enhances overall system reliability. Self-balancing occurs when detrimental factors counteract each other, thereby improving the system's ability to withstand adversity and recover. Symmetrical design often achieves self-balancing by eliminating undesired forces and moments in rotating machinery. For instance, symmetrical design can minimize axial forces on turbine shafts (Matthews, 1998). Symmetrical design also eliminates the need for thrust bearings in herringbone gear meshing. Increased torque transmission simultaneously increases the magnitude of each axial force in opposite directions, resulting in a minimal net axial force. Another example of self-balancing is twisting wires to cancel out magnetic interference. Twisted wires carry equal and opposite currents whose electromagnetic fields cancel each other out.

11.2.3 REDUCING THE VARIABILITY OF RISK-CRITICAL PARAMETERS

Reducing variability in critical risk parameters is essential to prevent them from reaching dangerous levels. Critical parameters encompass various types of variability, including: (i) variability associated with material properties, physical characteristics, manufacturing, and assembly; (ii) variability caused by product degradation over time; (iii) variability of operational loads experienced by the product; and (iv) variability caused by the operating environment.

Variability in strength due to production and property variations is a primary cause of the increased overlap between strength and load distributions, leading to failures from overstress. A pronounced lower tail in properties typically results in a similar tail in the strength distribution, thereby increasing the likelihood of early-life failures.

Choosing components from the same variety is a key strategy for minimizing variability. For mechanical components in contact, selecting both components with similar attributes (e.g., both with high surface hardness or both with normal surface hardness) is vital to reducing wear rates and preventing premature failure. Similarly, selecting transistors of the same variety to operate in parallel within a power supply circuit is critical for ensuring uniform distribution of the electrical loads among them, thereby avoiding overheating and premature failure. This principle applies equally to both electrical and mechanical components—for transistors in power circuits, resistors in sensitive bridge circuits, bearings in rotating shafts, and belts in drive systems.

In essence, maintaining uniformity in component selection is essential across various applications to achieve reliable performance and longevity.

Suppose that a particular key property (tolerance, strength, weight, etc.) is derived from pooling n batches, each containing m_i components. In each batch, the key property follows a particular unknown distribution with mean μ_i and standard deviation σ_i, $i = 1,\dots,n$. When all batches are combined into a single, large batch, the distribution of the key property in this single pooled batch becomes a mixture of n distributions, where $p_i = \dfrac{m_i}{\displaystyle\sum_{k=1}^{n} m_k}$ is the probability of selecting a component from the ith batch. The mean μ of the distribution of the key property in the pooled batch is given by

$$\mu = \sum_{i=1}^{n} p_i \mu_i \tag{11.3}$$

while the variance $V = \sigma^2$ of the property in the pooled batch is given by (Todinov, 2002b, 2003):

$$V = \sum_{i=1}^{n} p_i \left[\sigma_i^2 + \left(\mu_i - \mu \right)^2 \right] \tag{11.4}$$

As can be seen, the variance of the property in the pooled batch can be decomposed into two major components. The first component $\sum_{i=1}^{n} p_i \sigma_i^2$ characterises only the variation of properties within the batches while the second component $\sum_{i=1}^{n} p_i \left(\mu_i - \mu \right)^2$ of Equation (11.4) characterises the variation of properties between the separate batches. Assuming that all individual distributions have the same mean μ ($\mu_i = \mu_j = \mu$), the terms $p_i(\mu_i - \mu)^2$ in Equation (11.4), related to between-sources variation, become zero and the total variance becomes $V = \sum_{i=1}^{n} p_i \sigma_i^2$. In other words, in this case, the total variation of the property is entirely a within-sources variation (Todinov, 2003). This variation of properties can be reduced significantly if all components are selected from batch k, characterised by the smallest variance σ_k^2. In this case, selecting components from the same batch (the same variety of components) reduces the variance from $V = \sum_{i=1}^{n} p_i \left[\sigma_i^2 + \left(\mu_i - \mu \right)^2 \right]$ to $V = \sigma_k^2$.

Therefore, selecting components of the same variety is an effective method for reducing variability in properties and improving consistency. In some practical cases, the probability of selecting items of the same variety can be equated to the probability of assembling a reliable product.

11.3 ASSESSING THE PROBABILITY OF SELECTING ITEMS OF THE SAME VARIETY TO IMPROVE THE LEVEL OF BALANCING

Consider the common abstract inequality

$$a^2 + b^2 \geq 2ab \tag{11.5}$$

which holds true for any real numbers a and b because the inequality can be obtained directly from the obvious inequality $(a - b)^2 \geq 0$. Suppose that the numbers a and b are not both equal to zero.

If both sides of inequality (11.5) are divided by the positive quantity $(a + b)^2$, the next inequality

$$\frac{a^2}{(a+b)^2} + \frac{b^2}{(a+b)^2} \geq \frac{2ab}{(a+b)^2} \tag{11.6}$$

is obtained which can be easily reverse engineered (Todinov, 2022b).

If a and b represent the number of items of variety A and variety B, respectively, in a large batch, the left-hand side of inequality (11.6) represents the probability that two randomly selected items from the batch will both be of the same variety (either both of variety A or both of variety B) (Todinov, 2022b). For instance, the varieties could correspond to components manufactured by two different machine centres. It is assumed that the batch is sufficiently large so that the probability of the second item being of a particular variety is practically independent of the variety of the first item.

The right-hand side of inequality (11.6) represents the probability that two randomly selected items will be of different varieties $(A, B$ or $B, A)$. There are two possibilities for the selected items: they can either be of the same variety or of different varieties. Consequently, the sum of the probabilities of the complementary events—'the items are of the same variety' and 'the items are of different varieties'—is equal to one. According to inequality (11.6), the probability that the two selected items will be of different varieties is at most 0.5, meaning it is either less than or equal to 0.5, but never greater than 0.5.

This prediction is somewhat counterintuitive given the symmetry of the situations leading to the same-variety outcome and the different-variety outcome. Due to this symmetry, flipping two identical coins is equally likely to result in the same outcome on both coins (heads/heads or tails/tails) as it is to result in different outcomes on the coins (heads/tails or tails/heads).

The probabilities calculated from the left- and right-hand sides of inequality (11.6) can differ significantly. In an example considering 300 items of variety A and 800 items of variety B, the left-hand side of inequality (11.6) yields a probability of 0.603 for selecting two components of the same variety, and a probability of 0.397 for selecting two components of different varieties.

Interestingly, there is no analogous inequality related to selecting more than two components at random from a large batch containing two different varieties (Todinov, 2022b). Thus, the probability of selecting n components of the same variety ($n > 2$) from a large batch ($n < < a + b$) is given by

$$p_1 = a^n / (a+b)^n + b^n / (a+b)^n \qquad (11.7)$$

while the probability of not selecting n components of the same variety is $p_2 = 1 - p_1$ as a probability of a complementary event.

Denoting $u = a/(a + b)$ and substituting in (11.7) yields $p_1 = u^n + (1 - u)^n$. Let us conjecture that the inequality $p_1 \geq p_2$ holds for $n > 2$. This conjecture is equivalent to the conjectured inequality

$$u^n + (1-u)^n \geq 1/2 \qquad (11.8)$$

The maximum of the expression $x^n + y^n$, where $x + y = 1$, is obtained when $x = y = 1/2$. Consequently, the maximum of the left-hand side of (11.8) is obtained for $u = 1 - u$ ($u = 1/2$). However, $(1/2)^n + (1/2)^n < 1/2$ for any $n > 2$. The only value for which equality in (11.8) is attained is $n = 2$. As a result, inequality (11.8) does not hold for $n > 2$.

Now, let us conjecture that the inequality $p_1 \leq p_2$ holds for $n > 2$. This is equivalent to the inequality

$$u^n + (1-u)^n \leq 1/2 \qquad (11.9)$$

The conjectured inequality (11.9) can be disproved by providing a counterexample. For example, the inequality does not hold for $n = 3$ and $u = 0.9$. By taking u sufficiently close to unity, inequality (11.9) can also be disproved for any other $n > 3$.

Furthermore, no inequality similar to (11.6) holds for more than two varieties. Indeed, let a, b and c be the number of items of variety A, B and C, in a large batch of items.

The quantity

$$p = a^n / (a+b+c)^n + b^n / (a+b+c)^n + c^n / (a+b+c)^n \qquad (11.10)$$

expresses the probability p of selecting n components of the same variety ($n > 2$) from a large batch ($n < < a + b + c$) containing three varieties. Denoting $u = a/(a + b + c)$, $v = b/(a + b + c)$, $w = c/(a + b + c)$ and substituting in (11.10) yield $p = u^n + v^n$

$+ w^n$. Let us conjecture that the inequality $p \geq 1/2$ holds for $n \geq 2$ which is equivalent to the conjectured inequality

$$u^n + v^n + w^n \geq 1/2 \qquad (11.11)$$

Since $u + v + w = 1$, the maximum of the left-hand side of inequality (11.11) is obtained when $u = v = w = 1/3$. However, $(1/3)^n + (1/3)^n + (1/3)^n < 1/2$ for any $n \geq 2$. As a result, for two or more than two varieties, inequality (11.11) does not hold for any $n \geq 2$.

11.4 UPPER BOUND OF THE PROBABILITY OF SELECTING EACH COMPONENT FROM DIFFERENT VARIETY

Consider the algebraic inequality

$$n! x_1 x_2 \ldots x_n \leq n!/n^n \qquad (11.12)$$

where $0 \leq x_i \leq 1$, $\sum_{i=1}^{n} x_i = 1$. This inequality can be proved by using the arithmetic-mean geometric-mean inequality, according to which

$$\left(x_1 x_2 \ldots x_n \right)^{1/n} \leq \frac{x_1 + x_2 + \ldots + x_n}{n} \qquad (11.13)$$

Inequality (11.13) states that the geometric mean of n positive numbers $x_1, x_2, \ldots,$ x_n is always smaller than or equal to their arithmetic mean. Substituting $\sum_{i=1}^{n} x_i = 1$ into (11.13), raising the positive left- and right-hand sides to the power of 'n' and multiplying both sides by $n!$ gives inequality (11.12).

Inequality (11.12) has a useful physical interpretation (Todinov, 2022b). Suppose that there are n different varieties $(A_1, A_2, \ldots, A_n)$ of components in a large batch, with fractions $x_1, x_2, \ldots, x_n$ $\left(\sum_{i=1}^{n} x_i = 1 \right)$. Inequality (11.12) then gives the upper bound of the probability of selecting n components from different varieties because n different varieties can be selected in $n!$ distinct ways.

Consider the case corresponding to $n = 2$ in inequality (11.12):

$$2 x_1 x_2 \leq 2!/2^2 = 1/2 \qquad (11.14)$$

The variables x_1 and x_2 are physically interpreted as fractions of items from two varieties A and B in a large batch of items. Let a and b be the number of items of variety A and variety B in the batch. Let $x_1 = a/(a + b)$, denote the probability of random selection of an item of variety A and $x_2 = b/(a + b)$—the probability of random selection of an item of variety B.

Suppose that a working assembly can always be made if at least two of the selected components are of the same variety. In this case, it is important to determine the upper limit of the probability of selecting items from different varieties, as such a selection would make it impossible to create a functional assembly. From inequality (11.14), it follows that the probability of a faulty assembly resulting from selecting two components of different varieties can never exceed 0.5, regardless of the proportions of the different varieties in the batch. This confirms the conclusions drawn in Section (11.3).

This reasoning can be extended for 3 varieties in the batch. Let a, b and c be the number of items of variety A, B and C in a large batch of items. Let $x_1 = a/(a + b + c)$, denote the probability of selecting an item of variety A, $x_2 = b/(a + b + c)$—the probability of selecting an item of variety B and $x_3 = c/(a + b + c)$—the probability of selecting an item of variety C.

Suppose that a functional assembly requires selecting at least two items of the same variety. Therefore, it is important to determine the maximum possible probability of selecting three items from three different varieties, as such a selection would make it impossible to create a functional assembly.

For $n = 3$ inequality (11.12) results in:

$$3! \, x_1 x_2 x_3 \le 3!/\,3^3 = 2\,/\,9 \qquad (11.15)$$

which predicts that the probability of selecting two items from different varieties never exceeds 2/9, irrespective of the proportions of the separate varieties in the batch. For the probability p of selecting at least two items of the same variety, we have: $p \ge 1 - 2/9 = 7/9$.

For n components, for the probability of selecting all components from different varieties inequality (11.12) gives an upper bound of $n!/n^n$, irrespective of the proportions of the separate varieties in the batch.

11.5   LOWER BOUND OF THE PROBABILITY OF RELIABLE ASSEMBLY BY REVERSE ENGINEERING OF THE CHEBYSHEV'S SUM INEQUALITY

Consider the algebraic inequality (Todinov, 2022b)

$$x_1^n + x_2^n + \ldots + x_n^n \ge \frac{1}{n^{n-1}} \qquad (11.16)$$

where $0 \le x_i \le 1$, $\displaystyle\sum_{i=1}^{n} x_i = 1$.

Inequality (11.16) can be proved by reducing its complexity through the Chebyshev's sum inequality. Without loss of generality, it can be assumed that

$x_1 \geq x_2 \geq \ldots \geq x_n$. From the bsic properties of inequalities (see Section 2.1.1) it follows that $x_1^{n-1} \geq x_2^{n-1} \geq \ldots \geq x_n^{n-1}$. According to the Chebyshev's sum inequality in Chapter 2 (see Section 2.1.6):

$$\frac{x_1^n + x_2^n + \ldots + x_n^n}{n} \geq \frac{x_1^{n-1} + x_2^{n-1} + \ldots + x_n^{n-1}}{n} \times \frac{x_1 + x_2 + \ldots + x_n}{n}$$
$$= \frac{1}{n^2} \times \left(x_1^{n-1} + x_2^{n-1} + \ldots + x_n^{n-1} \right) \qquad (11.17)$$

As a result,

$$x_1^n + x_2^n + \ldots + x_n^n \geq \frac{1}{n^1} \times \left(x_1^{n-1} + x_2^{n-1} + \ldots + x_n^{n-1} \right) \qquad (11.18)$$

In exactly the same way, the complexity of $x_1^{n-1} + x_2^{n-1} + \ldots + x_n^{n-1}$ in the right-hand side of (11.18) can be reduced:

$$\frac{x_1^{n-1} + x_2^{n-1} + \ldots + x_n^{n-1}}{n} \geq \frac{x_1^{n-2} + x_2^{n-2} + \ldots + x_n^{n-2}}{n} \times \frac{x_1 + x_2 + \ldots + x_n}{n}$$
$$= \frac{1}{n^2} \times \left(x_1^{n-2} + x_2^{n-2} + \ldots + x_n^{n-2} \right)$$

As a result,

$$x_1^n + x_2^n + \ldots + x_n^n \geq \frac{1}{n^2} \times \left(x_1^{n-2} + x_2^{n-2} + \ldots + x_n^{n-2} \right) \qquad (11.19)$$

The complexity of $x_1^{n-2} + x_2^{n-2} + \ldots + x_n^{n-2}$ in the right-hand side of (11.19) can in turn be reduced by applying the Chebyshev's sum inequality and continuing this process finally leads to

$$x_1^n + x_2^n + \ldots + x_n^n \geq \frac{1}{n^{n-1}} \times \left(x_1^1 + x_2^1 + \ldots + x_n^1 \right) = \frac{1}{n^{n-1}} \qquad (11.20)$$

which completes the proof of inequality (11.16).

Let $x_1, x_2, \ldots, x_n$ be the fractions of components from varieties $A_1, A_2, \ldots, A_n$, respectively, in a large batch of components. The left-hand side of inequality (11.16) can be physically interpreted as the probability of selecting n components of the same variety and the right-hand side of the inequality is the lower bound of this probability.

The obtained lower bound of the probability of selecting components of the same variety can be used to determine the upper bound of the probability of not selecting components of the same variety.

Consider the complementary events A (all selected components are of the same variety) and $\bar{A}$ (not all selected components are of the same variety) whose union is the certain event Ω: $A \cup \bar{A} = \Omega$. Consequently, $P(A) + P(\bar{A}) = 1$. Assessing the upper bound U of the probability of event $\bar{A}$: $P(\bar{A}) \leq U$ can be made by using the lower bound L of the probability of the complementary event A:

$$P(\bar{A}) = 1 - P(A) \qquad (11.21)$$

Since for complementary events A and $\bar{A}$, the relationship (11.21) holds, it follows from (11.21) that the upper bound U of $P(\bar{A})$ is equal to 1 – the lower bound L of $P(A)$:

$$P(\bar{A}) \leq 1 - L \qquad (11.22)$$

As a result, the upper bound U of the probability $P(\bar{A})$ that not all components are of the same variety is

$$P(\bar{A}) \leq 1 - \frac{1}{n^{n-1}} \qquad (11.23)$$

Applying the principle of inversion in calculating the upper bound of the probability $P(\bar{A})$ through the lower bound L of the probability $P(A)$ significantly simplified models and calculations. Additional examples of using inversion to simplify models and calculations are discussed in (Todinov, 2019b).

This technique will be demonstrated with a large batch containing three varieties A_1, A_2 and A_3 with fractions x_1, x_2 and x_3. According to inequality (11.16), the lower bound of the probability of purchasing all three components of the same variety (event A) is given by

$$P(A) = x_1^3 + x_2^3 + x_3^3 > = 1/3^2 \qquad (11.24)$$

The probability $P(\bar{A})$ that not all components will be of the same variety is given by

$$P(\bar{A}) = 3x_1^2 x_2 + 3x_2^2 x_1 + 3x_2^2 x_3 + 3x_3^2 x_2 + 3x_3^2 x_1 + 3x_1^2 x_3 + 6x_1 x_2 x_3 \qquad (11.25)$$

which is a complex expression. This is a sum of the probabilities of: selecting two components of variety A_1 and one component of another variety, selecting two components of variety A_2 and one component of another variety, selecting two components of variety A_3 and one component of another variety, and selecting three components of three different varieties. As can be verified, $P(A) + P(\bar{A}) = 1$ because

$$x_1^3 + x_2^3 + x_3^3 + 3x_1^2 x_2 + 3x_2^2 x_1 + 3x_2^2 x_3 + 3x_3^2 x_2 + 3x_3^2 x_1 + 3x_1^2 x_3 + 6x_1 x_2 x_3$$
$$= (x_1 + x_2 + x_3)^3 = 1 \qquad (11.26)$$

According to Equation (11.23), the upper bound U of the probability $P(\bar{A})$ is given by

$$P(\bar{A}) \le 1 - 1/3^2 = 8/9 \qquad (11.27)$$

These results have been confirmed by Monte Carlo simulations.

11.6 REVERSE ENGINEERING OF ALGEBRAIC INEQUALITIES TO AVOID AN OVERESTIMATION OF EXPECTED PROFIT

11.6.1 AVOIDING THE RISK OF OVERESTIMATING PROFIT THROUGH REVERSE ENGINEERING OF THE JENSEN'S INEQUALITY

Consider the concave function $f(x)$ for which the Jensen's inequality holds:

$$f(w_1 x_1 + w_2 x_2 + \ldots + w_n x_n) \ge w_1 f(x_1) + w_2 f(x_2) + \ldots + w_n f(x_n) \quad (11.28)$$

where w_i ($i = 1, \ldots, n$) are weights that satisfy $0 \le w_i \le 1$ and $w_1 + w_2 + \ldots + w_n = 1$.

If the weights are chosen to be equal $w_i = 1/n$, the Jensen's inequality (11.28) becomes

$$f\left((1/n) \sum_{i=1}^{n} x_i \right) \ge (1/n) \sum_{i=1}^{n} f(x_i) \qquad (11.29)$$

The demand for a particular product X is almost always associated with variation (X is a random variable) and often, the variation of X cannot be controlled. The profits Y depend on the demand X through a particular function $Y = f(X)$. Now, suppose that the variables x_i in inequality (11.29) are interpreted as 'levels of demand' for a particular product and $f(x_i)$ are the profits corresponding to these levels of demand. The reverse engineering of inequality (11.29) then yields that the average of the profits at different levels of the demand is smaller than the profit calculated at an average level of the demand. The inequality helps the decision-maker choose between two competing strategies:

a. Averaging different values of the demand X, $\bar{x} = (1/n)\sum_{i=1}^{n} x_i$, by using n random values $x_1, x_2, \ldots, x_n$ of the demand within the demand range, followed by assessing the average profit from $\bar{y} = f(\bar{x})$;

b. Obtaining the average profit $\bar{y} = (1/n)\sum_{i=1}^{n} f(x_i)$ by averaging the profits corresponding to n random levels of demand: $x_1, x_2, \ldots, x_n$ within the demand range. Inequality (11.29) effectively states that strategy 'a' overestimates the profit.

To demonstrate the significant difference between the two decision strategies, consider an example from the chemical industry, where the demand x for a particular chemical product varies from 0 to 300000 kg per year and the capacity of the production plant for one year is only 200000 kg.

Suppose that the profit generated from selling the product, if the demand is in the interval (0,200000), is given by $f(x) = 3.6x$, where x is the quantity of the product [in kg] sold. The profit generated from selling the product if the demand is in the interval (200000,300000) is constant, given by $f(x) = 3.6 \times 200000$ because the production capacity cannot exceed 200000 kg.

The profit function is therefore a concave function (Figure 11.2), defined in the following way:

$$f(x) = \begin{cases} 3.6x, & 0 \leq x \leq 200000 \\ 3.6 \times 200000, & 200000 \leq x \leq 300000 \end{cases} \tag{11.30}$$

The average demand is obviously 300000/2 = 150000 kg per year. The profit corresponding to the average demand is $\bar{y}_1 = 3.6 \times 150000 = 540000$. This is the value $\bar{y}_1 = f\left((1/n)\sum_{i=1}^{n} x_i\right)$ on the left side of inequality (11.29).

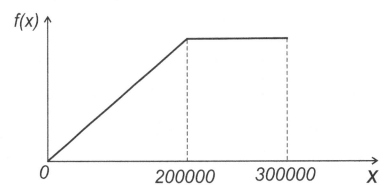

FIGURE 11.2   Concave profit-demand dependence generated from selling a product.

The average of the profits was calculated by using a simple Monte Carlo simulation. Running the Monte Carlo simulation with 100000 trials resulted in an average profit equal to 480000. This is the value $\bar{y}_2 = (1/n) \sum_{i=1}^{n} f(x_i)$ on the right-hand side of Inequality (11.29). The discrepancy between the two average profit values is considerable. Due to this significant difference, it is risky to base critical business decisions on profits calculated at average demand. To avoid the risk of over-estimating average profit, a more realistic strategy is to calculate the average of profits at various demand levels, rather than estimating the profit at an average demand level.

11.6.2 AVOIDING OVERESTIMATION OF THE AVERAGE PROFIT THROUGH REVERSE ENGINEERING OF THE CHEBYSHEV'S SUM INEQUALITY

Avoiding the overestimation of the average profit can also be illustrated with the reverse engineering of the Chebyshev's sum inequality, introduced in Chapter 2 (Section 2.1.6).

Let $a_1, \ldots, a_n$ and $b_1, \ldots, b_n$ be two sets of similarly ordered positive numbers $(a_1 \leq, \ldots, \leq a_n$ and $b_1 \leq, \ldots, \leq b_n$. Then the Chebyshev's sum inequality holds:

$$\frac{a_1 b_1 + \ldots + a_n b_n}{n} \geq \frac{a_1 + \ldots + a_n}{n} \cdot \frac{b_1 + \ldots + b_n}{n} \qquad (11.31)$$

If the number sequences are oppositely ordered, for example $a_1 \geq, \ldots, \geq a_n$ and $b_1 \leq, \ldots, \leq b_n$ hold, the inequality is reversed.

$$\frac{a_1 b_1 + \ldots + a_n b_n}{n} \leq \frac{a_1 + \ldots + a_n}{n} \cdot \frac{b_1 + \ldots + b_n}{n} \qquad (11.32)$$

Suppose that in inequality (11.32), the variables a_i ($i = 1, \ldots, n$), ranked in descending order ($a_1 \geq, \ldots, \geq a_n$) stand for probabilities of a gain from risky investments. The variables b_i ($i = 1, \ldots, n$), ranked in ascending order ($b_1 \leq, \ldots, \leq b_n$), stand for the returns from the risky investments which correspond to the probabilities a_i. This negative correlation between the probabilities of a gain and the returns from risky investments is quite common: the lower the likelihood of success in a risky investment, the higher the magnitude of the potential return.

The right-hand side of inequality (11.32) can now be physically interpreted as an estimate of the expected profit per investment by using the average return from the investments $\bar{b} = (b_1 + b_2 + \ldots + b_n)/n$ and the average probability of a gain $\bar{a} = (a_1 + \ldots + a_n)/n$. The left-hand side of inequality (11.32) can be physically interpreted as the expected profit per investment, assessed by taking the average of the expected profits from the individual investments. Inequality (11.32), predicts that estimating the expected profit from investments by using the average return and the average probability of a gain, leads to an overestimation of the expected profit per investment.

The overestimation of the expected profit per investment can be significant and this will be illustrated by a very simple numerical example involving only two investments: an investment with return $c_1 = \$800$ and an investment with return $c_2 = \$15000$. The probabilities of a gain from the investments are $p_1 = 0.77$ and $p_2 = 0.60$, correspondingly. If the average probability of return from an investment and the average return are used, the value $\bar{p} \times \bar{c} = 0.5 \times (0.77 + 0.60) \times 0.5 \times (800 + 15000) = 5411$ is obtained for the expected potential profit per investment. The actual expected potential profit per investment is

$$\frac{p_1 c_1 + p_2 c_2}{2} = \frac{0.77 \times 800 + 0.60 \times 15000}{2} = 4808$$

which is significantly smaller than the prediction of 5411, based on the average values of the probabilities and the returns.

In the Chebyshev's sum inequality (11.32), the sequences $a_1 \geq, ..., \geq a_n$ could also stand for interest rates corresponding to each of n customer loans represented by the variables b_i $(i = 1, ..., n)$, ranked in ascending order: $b_1 \leq, ..., \leq b_n$. The negatively correlated sequences $a_1 \geq, ..., \geq a_n$ and $b_1 \leq, ..., \leq b_n$ are common for lending agencies that profit from lending to customers. The interest charged on loans is often inversely related to the size of the customer's loan. Customers with smaller loans typically have poorer credit ratings, resulting in higher charged interest rates. Conversely, customers with larger loans generally have better credit ratings, leading to lower charged interest rates. The right-hand side of inequality (11.32) can then be physically interpreted as the projected expected profit $\overline{ab}$ for the lending agency per customer, made by using the average size of the loans $\bar{b} = (b_1 + b_2 + ... + b_n)/n$ and the average interest rate $\bar{a} = (a_1 + ... + a_n)/n$. The left-hand side of inequality (11.32) can be interpreted as the actual average profit for the lending agency per customer. The Chebyshev's inequality (11.32), now predicts that the projected expected profit per customer, estimated by using the average loan $\bar{b} = (b_1 + b_2 + ... + b_n)/n$ and the average charged interest rate $\bar{a} = (1/n)(a_1 + ... + a_n)$, is overestimated.

The overestimation of the projected expected profit per customer can be significant and this will be illustrated by a simple numerical example involving two customers only, with loans $b_1 = \$600$ and $b_2 = \$40000$. Suppose that the interest rates charged are $a_1 = 0.12$ and $a_2 = 0.04$ correspondingly. If the average interest rate and the average loan are used for calculating the projected average profit per customer, the value $\bar{a} \times \bar{b} = 0.5 \times (0.12 + 0.04) \times 0.5 \times (600 + 40000) = 1624$ will be predicted.

However, the actual average profit per customer is $\dfrac{p_1 b_1 + p_2 b_2}{2} = \dfrac{0.12 \times 600 + 0.04 \times 40000}{2} = \$836$, which is significantly smaller than the predicted value.

The conclusion is that for negatively correlated variables, calculating the average of their dot product by multiplying the averages of the two variables leads to an overestimation.

Conversely, according to Inequality (11.31), for positively correlated variables, calculating the average of their dot product by multiplying the averages of the two variables leads to an underestimation.

11.7 AVOIDING OVERESTIMATION OF THE PROBABILITY OF SUCCESSFUL ACCOMPLISHMENT OF MULTIPLE TASKS

Consider inequality 2.9 proved in Section 2.1.3 of Chapter 2:

$$\left(\frac{m_1 x_1 + m_2 x_2 + \ldots + m_n x_n}{M} \right)^M \geq x_1^{m_1} x_2^{m_2} \ldots x_n^{m_n} \tag{11.33}$$

where $m_1, \ldots, m_n$ and $x_1, \ldots, x_n$ are positive values, and $M = \sum_{i=1}^{n} m_i$.

Let x_i ($0 < x_i < 1$) stand for the probability of successfully completing a task of type i, where $i = 1, 2, \ldots, n$. Successfully completing any particular task is statistically independent of successfully completing any other task. Let m_i stand for the number of tasks of type i where $i = 1, 2, \ldots, n$.

The right-hand side of inequality (11.33) can be physically interpreted as the probability of successfully completing all $M = \sum_{i=1}^{n} m_i$ tasks.

Indeed, the probability of successfully completing m_1 tasks of type one is $x_1^{m_1}$; the probability of successfully completing m_2 tasks of type two is $x_2^{m_2}$ and so on. The probability of successfully completing all n types of tasks is, therefore, given by the right-hand side of inequality (11.33).

The expression $\bar{x} = \frac{m_1 x_1 + m_2 x_2 + \ldots + m_n x_n}{M}$ on the left-hand side of inequality (11.33) can be physically interpreted as the average probability $\bar{x}$ of successfully completing a task, irrespective of its type. It is simply obtained by adding the probabilities characterising the separate tasks and dividing the sum by the total number $M = \sum_{i=1}^{n} m_i$ of tasks. Inequality (11.33) can be rewritten as

$$\bar{x}^M \geq x_1^{m_1} x_2^{m_2} \ldots x_n^{m_n} \tag{11.34}$$

Inequality indicates that the calculated probability of successfully completing all M tasks, based on the average probability of completing a single task, is higher than the actual probability of successfully completing all M tasks. The reverse engineering of inequality (11.33) revealed that using the average probability of completing a task results in an overestimation of the actual probability of successfully completing all tasks.

12 Generating New Knowledge by Reverse Engineering of Algebraic Inequalities in Terms of Potential Energy

12.1 REVERSE ENGINEERING OF AN INEQUALITY IN TERMS OF POTENTIAL ENERGY

Consider the general algebraic inequality

$$f\left(x_1, x_2, \ldots, x_n\right) \geq L \tag{12.1}$$

where x_1, x_2, ..., x_n are system/process parameters and L is an unknown lower bound. The parameters may be subjected to a constraint:

$$\varphi\left(x_1, x_2, \ldots, x_n\right) = 0$$

where $\varphi(x_1, x_2, \ldots, x_n) = 0$ is an arbitrary continuous function of the system/process parameters x_1, x_2, ..., x_n. A common constraint is $\varphi \equiv \sum_{i=1}^{n} x_i = d$, where d is a constant.

Often, the function $f(x_1, x_2, \ldots, x_n)$ can be interpreted as *potential energy* of a system (Todinov, 2021b). If this can be done, the constant L in the right-hand of inequality (12.1) can be interpreted as the minimum potential energy of the system which corresponds to its state of stable equilibrium. From the equilibrium conditions that correspond to the stable equilibrium, a number of useful relationships can be derived and the lower bound L determined without resorting to complex models. To illustrate the idea behind the potential energy interpretation of algebraic inequalities (Todinov, 2021b), a simple example will be used. Consider the inequality:

$$\sqrt{a^2 + x^2} + \sqrt{b^2 + y^2} \geq L \tag{12.2}$$

DOI: 10.1201/9781003517764-12

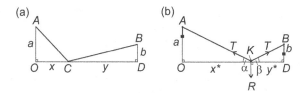

FIGURE 12.1 Meaningful interpretation of inequality (12.1) in terms of potential energy of a constant-tension spring: (a) potential energy corresponding to an arbitrary location of point C; (b) location of the ring K which corresponds to a minimal potential energy of the system.

where a and b are constants and the non-negative parameters x, y are subjected to the constraint

$$x + y = d$$

where $d = OD$ (Figure 12.1a). In inequality (12.2), the lower bound L is unknown quantity.

Noticing that both $\sqrt{a^2 + x^2}$ and $\sqrt{b^2 + y^2}$ can be interpreted as a hypotenuse of a right-angle triangle, inequality (12.2) can be physically interpreted by using the two right-angle triangles OAC and CBD in Figure 12.1a.

The left-hand side of inequality (12.2) is the length of the path ACB and the lower bound L is the smallest value of the length ACB. In interpreting the left-hand side of inequalities (12.1) and (12.2) as potential energy, physical analogies such as *constant tension springs*, *zero-length linear springs* and *zero-length non-linear springs* prove to be useful.

A *constant-tension* spring is a spring whose tension is independent of its length (Levi, 2009). It is assumed that the tension T in the constant-tension spring when the system is in equilibrium, is equal to unity ($T = 1$).

The potential energy E of a constant-tension spring of length x is given by

$$E = \int_0^x T\,du = Tx = x \qquad (12.3)$$

which is product of the spring length x and the tension of the spring T ($T = 1$). Note that the potential energy of a constant-tension string ($E = x$) as a function of the spring length x is different from the potential energy $E = (1/2)kx^2$ of a conventional spring stretched to a displacement x. This is because, for a conventional spring, the spring force $F = kx$ varies proportionally with the displacement x, while for a constant-tension spring the spring force is constant $F = 1$ and does not depend on the displacement x. A constant-tension spring can be created using unit weights and pulleys (see Figure 12.1b). When the weights remain constant, the tension force T remains unchanged. However, the potential energy of the system varies with the spring length. As the length of the constant-tension spring $KA+KB$ increases (Figure 12.1b), the sum of the elevations of the suspended unit weights also increases, resulting in greater potential energy for the system.

Suppose that K in Figure 12.1b is a ring that moves without any friction along the segment OD, whose length is equal to d and AKB is a non-stretchable string kept under constant tension T through the pulleys A and B and the unit weights at the ends. The left part of inequality (12.2) is then proportional to the potential energy of the system in Figure 12.1b. The smallest potential energy of the system in Figure 12.1b corresponds to the smallest length $KA + KB$ of the string, which marks the smallest combined elevation of the unit weights.

The constant L on the right-hand side of inequality (12.2) represents the minimum potential energy of the system. For a state of stable equilibrium, the forces applied to point K on the string must balance out. Because the ring was assumed to be frictionless, the force R applied to the string from the ring is perpendicular to the x-axis and its component along the horizontal axis is zero. Consequently, the sum of the components of the tension force T applied on the string at point K, along the horizontal axis, must be equal to zero ($-T\cos\alpha + T\cos\beta = 0$). From this necessary condition for a system equilibrium, it is clear that the angle AKO (α) must be equal to angle BKD (β). In this case, triangles AKO and BKD are similar and $x^*/a = (d - x^*)/b$. From this relationship, $x^* = ad/(a + b)$ and $y^* = bd/(a + b)$. The value of the lower bound L in inequality (12.2) is equal to the smallest length $L = \sqrt{(a+b)^2 + d^2}$ of AKB. As a result, the exact value of the lower bound L in inequality (12.2) has been determined solely on the basis of the reverse engineering of inequality (12.2) in terms of potential energy of a constant-tension spring.

12.2 A NECESSARY CONDITION FOR MINIMISING A SUM OF POWERS OF DISTANCES

Consider n points in space: $A_1, A_2,\ldots,A_n$ and an extra point M with distances $r_1, r_2,\ldots,r_n$ to the points $A_1, A_2,\ldots,A_n$, respectively (Figure 12.2a). Consider the abstract inequality (12.4) involving the distances $r_1, r_2,\ldots,r_n$ (Todinov, 2021b):

$$r_1^n + r_2^n +\ldots+ r_n^n \geq L \tag{12.4}$$

where L is the lower bound which is unknown quantity.

Suppose also that the connecting segments $r_i = M_0 A_i$ (Figure 12.2b) are *nonlinear zero-length tension springs* of order $n - 1$. A nonlinear zero-length spring of order $n - 1$ is a spring whose tension T is directly proportional to the $(n - 1)$-th power of its length r ($T = kr^{n-1}$). The potential energy E of a stretched to a length r non-linear, zero-length spring of order $n - 1$, is given by

$$E = \int_0^r kv^{n-1}dv = kr^n/n \tag{12.5}$$

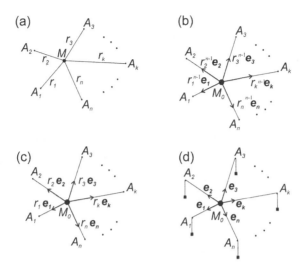

FIGURE 12.2 (a) Distances MA_i from point M to a specified set of points in space (b) the necessary condition for a minimal sum of distances M_0A_i raised to the power of n is vectors $r_i^{n-1}e_i$ to add up to zero; (c) the necessary condition for a minimal sum of squared distances M_0A_i is point M_0 to be located at the mass center of the system of points; (d) the necessary condition for a minimal sum of distances M_0A_i is point M_0 to be located at the geometric median of the system of points.

From Equation (12.5) it can be seen that if the constant k is set to be equal to n, the left-hand part of inequality (12.4) can be physically interpreted as the total potential energy of a system of n non-linear zero-length springs of order $n - 1$, while the right-hand part of inequality (12.4) can be physically interpreted as the minimum possible potential energy of the system of non-linear zero-length springs of order $n - 1$.

If point M is now selected at such a location (point M_0 in Figure 12.2b) that the sum of the nth power of the distances $r_1, r_2, \ldots, r_n$ is a minimum, the equality

$$r_1^n + r_2^n + \ldots + r_n^n = L \tag{12.6}$$

will hold. Since the minimum potential energy corresponds to a stable equilibrium of the system, the sum of the vectors $r_i^{n-1}\mathbf{e_i}$ with origin point M_0, directed towards the separate points $A_1, A_2, \ldots, A_n$, must be zero (Figure 12.2b):

$$\sum_{i=1}^{n} r_i^{n-1}\mathbf{e_i} = 0 \tag{12.7}$$

Figure 12.2b demonstrates that the physical interpretation of the left- and right-hand part of inequality (12.4) in terms of potential energy of a system of non-linear springs leads to a necessary condition for determining the optimal location M_0. The point M_0 that corresponds to the minimal sum of distances to the specified points A_i, raised to power of n, coincides with the point for which $\sum_{i=1}^{n} r_i^{n-1} \mathbf{e_i} = 0$. There are two important special cases of inequality (12.4) which correspond to the cases $n = 2$ and $n = 1$.

12.3 DETERMINING THE LOWER BOUND OF THE SUM OF SQUARED DISTANCES TO A SPECIFIED NUMBER OF POINTS IN SPACE

Consider the special case of inequality (12.4) where $n = 2$. In this case, inequality (12.4) transforms into the inequality

$$r_1^2 + r_2^2 + \ldots + r_n^2 \geq L \tag{12.8}$$

The connecting segments $r_i = M_0 A_i$ in Figure 12.2c can be physically interpreted as *zero-length linear tension springs*. The tension T of a linear zero-length tension spring is directly proportional to its length r ($T = kr$). According to equation (12.5), the potential energy of a stretched to a length r linear spring is given by $(1/2) k r^2$. If the spring stiffness k is selected to be equal to 2 ($k = 2$), the left-hand part of inequality (12.8) is the potential energy of a system of n zero-length linear springs while the right-hand part of inequality (12.8) is the minimum potential energy of the system of springs. Since the minimum total potential energy corresponds to a stable equilibrium of the system, the sum of the vectors $r_i \mathbf{e_i}$ from point M_0 directed towards the separate points $A_1, A_2, \ldots, A_n$ must be zero: $\sum_{i=1}^{n} r_i \mathbf{e_i} = 0$ (see Figure 11.2c). This means that, in this case, point M_0 must coincide with the mass centre of the system of points.

As shown in Figure 11.2c, the physical interpretation of the left- and right-hand sides of inequality (12.4) for $n = 2$ led to a necessary condition for the location of point M_0. This point, which minimizes the sum of squared distances to a set of fixed points in space, coincides with the centroid of the system of points.

12.4 A NECESSARY CONDITION FOR DETERMINING THE LOWER BOUND OF A SUM OF DISTANCES

Consider now a special case of the inequality (12.4), where $n = 1$. In this case, inequality (12.4) transforms into the inequality (Todinov, 2021b):

$$r_1 + r_2 + \ldots + r_n \geq L \tag{12.9}$$

Inequality (12.9) can also be physically interpreted in a meaningful way. Suppose that the connecting segments $M_0 A_i$ are constant-tension springs (Figure 12.2d).

Assume that the tension T in each of these strings, when the system is in equilibrium, is equal to unity ($T = 1$).

According to equation (12.3), the potential energy of a stretched constant-tension spring to a length r is given by kr, which is product of the spring length and the tension of the spring ($k = T = 1$). The constant-tension string shown in Figure 12.2d is constructed using unit weights and pulleys. The left-hand side of inequality (12.9) is proportional to the system's potential energy in Figure 12.2d. As the sum of the distances M_0A_i to the individual points increases, so does the sum of the elevations of the suspended unit masses, resulting in higher potential energy for the system. The system's potential energy reaches a minimum when the sum of the distances M_0A_i to the points is minimized (Figure 12.2d).

Since the minimum potential energy corresponds to a stable equilibrium of the system, the sum of the unit vectors e_i from point M_0 directed towards the separate points must be zero: $\sum_{i=1}^{n} e_i = 0$ (see Figure 12.2d). As aresult, the interpretation of the left- and right-hand sides of inequality (12.9) provides a necessary condition for determining the optimal location of point M_0. The point with the minimal sum of distances to the points $A_1, A_2, \ldots, A_n$ is where the sum of the *line-of-sight* unit vectors towards the points is zero: $\sum_{i=1}^{n} e_i = 0$. In this case, the optimal location M_0 coincides with the *geometric median* (Wesolowsky, 1993) of the system of points. Finding the location M_0 that minimises the sum of the distances to the specified points is the famous *Fermat-Weber location problem* (Chandrasekaran and Tamir, 1990). Algorithms for determining the location M_0 have been presented in Weiszfeld (1937, Chandrasekaran and Tamir (1990) and Papadimitriou and Yannakakis (1982).

The described approach to minimising the sum of distances can be applied not only for a set of points. In Figure 12.3 the circle λ with radius r and coordinates (d,b) of the centre C ($d = OD$, $b = CD$) is inside the angle DOE whose size will be denoted by γ. It is required to select a point K on OD (the x-axis) such that AKB, which is the sum of the distance KA to OE and the distance KB to the circle λ, is minimal. It is not difficult to observe that the smallest length AKB corresponds to

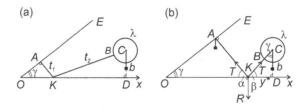

FIGURE 12.3 Minimising the sum of distances of point K to OE and the circle λ: (a) an arbitrary location of point K; (b) location of point K which corresponds to the minimal potential energy of the system.

the smallest length AKC. Therefore, AKB is minimised whenever AKC is minimised. The task consists of determining the lower bound L of

$$AK + KC \geq L \qquad (12.10)$$

Suppose again that in Figure 12.3b, K, is a ring that moves without any friction along the segment OD whose length is equal to d and AKC is a non-stretchable string kept under constant tension by the pulleys A and C and the unit weights at the ends.

The sum AKC is then proportional to the potential energy of the system in Figure 12.3b. The smallest potential energy of the system in Figure 12.3b corresponds to the smallest length $KA + KC$ of the string which marks the smallest combined elevation of the unit weights. This is also the state of equilibrium of the system. From the necessary condition for equilibrium of the ring K, it is clear that the angle CKD must be equal to angle AKO. From the right-angle triangles OAK and KCD, it follows that the angle KCD must be equal to γ. As a result, $KC = b/\cos \gamma$, $KB = b/\cos \gamma - r$ and the optimal distance DK is found from $DK = b \tan (\gamma)$. Since $OK = d - b \tan \gamma$, $AK = (d - b \tan \gamma) \sin \gamma$, the minimal length AKB is obtained from $AKB = b/\cos \gamma - r + (d - b \tan \gamma) \sin \gamma$. The physical interpretation in terms of potential energy significantly simplified the solution.

12.5   A NECESSARY CONDITION FOR DETERMINING THE LOWER BOUND OF THE SUM OF SQUARES OF TWO QUANTITIES

A simple but an important illustration of the potential energy interpretation can be given with the inequality

$$a^2 + b^2 \geq L \qquad (12.11)$$

where a and b are two non-negative quantities whose sum is equal to a given constant d: $d = a + b$ and L is the lower bound which is unknown quantity.

Suppose that in Figure 12.4, $PD = a$ and $DE = b$ are linear zero-length springs. P is a ring that moves without any friction along AC which has a length of $d = a + b$ while E is a ring which moves without any friction along BC, also with a length of $d = a + b$. Similarly, D is a ring that moves without any friction along AB. The segment AB is a hypotenuse of the isosceles right-angle triangle ABC and the angles EDB and PDA remain constantly equal to $\pi/4$.

The left-hand side of inequality (12.11) is equal to the total potential energy of the system of linear springs in Figure 12.4a. The smallest potential energy of the system in Figure 12.4b corresponds to the equilibrium position of the rings P,E and D. At the equilibrium position of ring D, the sum of the projections of forces F_1 and F_2 acting on the ring D must be equal to zero ($-F_1 \cos (\pi/4) + F_2 \cos (\pi/4) = 0$) which is only possible if $F_1 = F_2$.

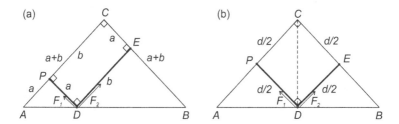

FIGURE 12.4 (a) Squared sum $a^2 + b^2$ which corresponds to an arbitrary location of point D and (b) location of point D which corresponds to the minimal squared sum $a^2 + b^2$.

Since the linear spring forces are proportional to the spring lengths ($F_1 = ka$; $F_2 = kb$), this means that if the system is in equilibrium, the ring D must be in the middle of segment AB. In this position, DP = DE = $d/2$, $|F_1| = |F_2|$ and for $k = 1$, the minimum potential energy E_{min} of the system becomes

$$E_{min} = (1/2)(d/2)^2 + (1/2)(d/2)^2 = d^2/4 = (a+b)^2/4$$

Since the potential energy of the system in Figure 12.4a at an arbitrary location of the ring D is $a^2/2 + b^2/2$, the following inequality holds

$$a^2/2 + b^2/2 \geq E_{min} = d^2/4 \qquad (12.12)$$

Multiplying inequality (12.12) by two gives

$$a^2 + b^2 \geq d^2/2 \qquad (12.13)$$

Consequently, for the constant L in inequality (12.11), $L = (a + b)^2/2$ has been determined. The lower bound L in inequality (12.11) is half of the squared sum of the quantities a and b. The exact value of the lower bound L in inequality (12.11) was derived purely from a meaningful physical interpretation of inequality (12.11) in terms of potential energy.

12.6   REVERSE ENGINEERING OF A GENERAL INEQUALITY INVOLVING A MONOTONIC CONVEX FUNCTION

The discussion presented in the previous sections related to determining a lower bound by physically interpreting inequalities as potential energy of springs can be extended in the general case of non-linear springs whose potential energy is defined by a general function $f(x)$. The constraints on the function $f(x)$ are that it must be non-negative, strictly increasing or strictly decreasing within a specified interval $0 \leq x \leq d$, and convex in that interval. A convex function is one that is differentiable within the interval $0 \leq x \leq d$ and has a second derivative greater than zero throughout that interval ($f''(x) > 0$, $0 \leq x \leq d$).

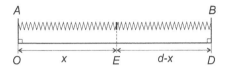

FIGURE 12.5 Physical interpretation of inequality (12.13) in terms of total potential energy of two non-linear springs.

Consider the inequality

$$f(x) + f(y) \geq L \tag{12.13}$$

where, $0 \leq x \leq d$, $0 \leq y \leq d$ and $x + y = d$ are constraints and L is unknown lower bound.

It will be shown that the reverse engineering of inequality (12.13) in terms of potential energy of a system of springs yields that the lower bound L is given by $L = 2f(d/2)$.

Consider two non-linear springs AE and BE whose potential energy as a function of their length is given by $U = f(x)$ (Figure 12.5). For the segment OD, we have OD = d. The total potential energy $T(x)$ of the system of springs in Figure 12.5, as a function of the distance x of point E from the origin O of the coordinate system, is given by the left-hand side of inequality (12.13)

$$T(x) = f(x) + f(d - x) \tag{12.14}$$

where $f(x) \geq 0$ for $x \geq 0$.

Differentiating $T(x)$ with respect to x gives

$$T'(x) = f'(x) - f'(d - x) \tag{12.15}$$

At $x = d/2$, $T'(x) = 0$ therefore point $x = 0$ is a stationary point. The second derivative of the total potential energy is given by $T''(x) = f''(x) + f''(d - x)$. Since $f''(x) > 0$ by the choice of $f(x)$ as a convex function, it follows that $T''(x) > 0$; therefore, the stationary point corresponds to a local minimum of the potential energy.

The local minimum is also a global minimum in the interval $0 \leq x \leq d$. This is because the total potential energy takes equal values at the boundaries of the domain ($T(0) = T(d) = f(0) + f(d)$). Additionally, the first derivative of the potential energy $T(x)$, given by equation (12.15,) can be zero only at $x = d/2$, indicating a single stationary point in the interval $[0,d]$.

The spring force is a conservative force. Its work depends solely on the length of the spring, not on the path taken to reach the final length of the spring or the time involved. Consequently, the coordinates x, y and z of the end of a stretched spring

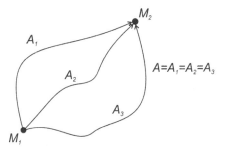

FIGURE 12.6 Conservative stationary field.

define a conservative stationary field $\Phi(x, y, z)$. The work A in a conservative stationary field $\Phi(x, y, z)$ depends only on the initial M_1 and final position M_2 in the field but not on the trajectory (Figure 12.6).

The work A done in the stationary filed $\Phi(x, y, z)$ for moving an object from M_1 to M_2 is given by the line integral

$$A = \int_{M_1}^{M_2} \left(F_x dx + F_y dy + F_z dz \right) \qquad (12.16)$$

where F_x, F_y, F_z are the components along the corresponding coordinate axes, of the force $\mathbf{F}$ acting on the material point. The necessary and sufficient conditions for a stationary potential field are

$$F_z = \frac{\partial \Phi}{\partial z}; F_y = \frac{\partial \Phi}{\partial y}; F_z = \frac{\partial \Phi}{\partial z} \qquad (12.17)$$

As a result, the elementary work done in a potential filed is equal to the total differential of the potential filed:

$$dA = \frac{\partial \Phi}{\partial x} dx + \frac{\partial \Phi}{\partial y} dy + \frac{\partial \Phi}{\partial z} dz = d\Phi \qquad (12.18)$$

and the total work done by the potential field for moving a material point from M_1 to M_2 is given by

$$A = \int_{M_1}^{M_2} d\Phi(x, y, z) = \Phi_2 - \Phi_1 \qquad (12.19)$$

As a result, the work required to move a material point from one location to another depends on the difference in potential field values between the initial and final points. From equation (12.19) it also follows that the work around a closed path in the potential field is zero:

$$A = \int_{M_1}^{M_1} d\Phi\left(x, y, z\right) = \Phi_1 - \Phi_1 = 0 \qquad (12.20)$$

The potential energy U is the work done by the conservative field for moving the material point from M_1 to M_2, taken with a negative sign. It expresses the storage of energy at a particular location in the conservative field.

Considering the definition of potential energy, the components of the force **F** acting on the material point become $F_x = -\dfrac{\partial U}{\partial x}$, $F_y = -\dfrac{\partial U}{\partial y}$, $F_z = -\dfrac{\partial U}{\partial z}$

Going back to the system of springs in Figure 12.5, the forces F_1 and F_2 with which the springs act on the junction point E (Figure 12.5) are given by the derivatives of the potential energy of the spring:

$$F_1 = -\frac{df\left(x\right)}{dx}; F_2 = -\frac{df\left(y\right)}{dy}$$

At the equilibrium point, $\left|F_1\right| = \left|F_2\right|$.

For a convex function such as $f(x) = ke^{-x}$, where k is a constant, the inequality

$$ke^{-x} + ke^{-y} \geq 2ke^{-d/2} \qquad (12.21)$$

where $x + y = d$, holds. The global minimum of the potential energy $T(x) = ke^{-x} + ke^{-(d-x)}$ is attained in the middle of the interval $[0,d]$, for $x = d/2$, and is equal to $2ke^{-d/2}$.

For potential energy given by the convex function $f(x) = k/(x + r)$, where k and r are constants, the inequality

$$k/\left(x+r\right) + k/\left(y+r\right) \geq 2k/\left(d/2+r\right) \qquad (12.22)$$

where $x + y = d$ holds. The global minimum of the potential energy $T(x) = k/(x + r) + k/(y + r)$ (the lower bound) is attained in the middle of the interval $[0,d]$ for $x = d/2$ and is equal to $2k/(d/2 + r)$.

It is important to note that the minimum of the total potential energy is not always reached at the midpoint of the interval $[0,d]$ for any potential energy function $f(x)$. In particular, for the concave function $f\left(x\right) = \sqrt{x+1}$, this is definitely not the case. The inequality

$$\sqrt{x+1} + \sqrt{y+1} \geq 2\sqrt{d/2+1} \qquad (12.23)$$

where $x + y = d$ is not true. The minimum of the potential energy is obtained at the ends of the interval $[0, d]$ (for $x = 0$ or $x = d$) and is equal to $1 + \sqrt{d + 1}$. The correct inequality is

$$\sqrt{x+1} + \sqrt{y+1} \geq 1 + \sqrt{d+1} \tag{12.24}$$

Despite that the spring forces $F_1 = -\dfrac{d\left(\sqrt{x+1}\right)}{dx}$ and $F_2 = -\dfrac{d\left(\sqrt{y+1}\right)}{dy}$ are equal in the middle, at point E (Figure 12.5), point E corresponds to a local maximum and the equilibrium at this point is unstable. A slight displacement from point E leads to a decrease of the total potential energy.

The developments detailed in this book can be further expanded by generating new insights through reverse engineering of algebraic inequalities across diverse disciplines, including engineering, physics, chemistry, biology, medical research, environmental research, economics, management, and operational research. Notably, inequalities involving sub-additive and super-additive functions hold significant potential. These types of inequalities are especially amenable to reverse engineering when their variables and terms represent additive quantities.

References

Alsina C. and Nelsen R.B. (2010). *Charming Proofs: A Journey into Elegant Mathematics*. Washington, DC: The Mathematical Association of America.

Altshuller G.S. (1999). *The Innovation Algorithm, TRIZ, Systematic Innovation and Technical Creativity*. Worcester, MA: Technical Innovation Center, Inc.

Andriani P. and McKelvey B. (2007). Beyond Gaussian averages: Redirecting international business and management research toward extreme events and power laws. *Journal of International Business studies*, 38, 1212–1230.

Ang A.H. and Tang W.H. (2007). *Probability Concepts in Engineering: Emphasis on Applications in Civil and Environmental Engineering*, Hoboken, NJ: John Wiley & Sons.

Aqel I. and Mohamed Mellal M. (2023). Optimal reliability allocation of heterogeneous components in pharmaceutical production plant. *International Journal on Interactive Design and Manufacturing (IJIDeM)*, 17, 1711–1720.

Aven T. (2003). *Foundations of Risk Analysis*, Chichester: Wiley.

Aven T. (2017). Improving the foundation and practice of reliability engineering. *Proceedings of the Institution of Mechanical Engineers, Part O: Journal of Risk and Reliability*, 231(3), 295–305.

Bazovsky I. (1961). *Reliability Theory and Practice*, Englewood Cliffs, NJ: Prentice-Hall.

Bechenbach E. and Bellman R. (1961). *An Introduction to Inequalities*, New York: Random House, The L.W.Singer Company.

Beer F.P., Russell Johnston E. and DeWolf J.T. (2002). *Mechanics of Materials*, 3rd ed. New York: McGrawHill.

Ben-Haim Y. (2005). Info-gap Decision Theory for Engineering Design. Or: Why 'Good' is Preferable to 'Best', in *Engineering Design Reliability Handbook*, Edited by E. Nikolaidis, D. M. Ghiocel and S. Singhal, Boca Raton, FL: CRC Press.

Berg V.D. and Kesten H. (1985). Inequalities with applications to percolation and reliability. *Journal of Applied Probability*, 22 (3), 556–569.

Besenyei A. (2018). Picard's weighty proof of Chebyshev's sum inequality. *Mathematics Magazine*, 91(5), 366–371, DOI: 10.1080/0025570X.2018.1512814

Budynas R.G. (1999). *Advanced Strength and Applied Stress Analysis*, 2nd ed. New York: McGraw-Hill.

Budynas R.G. and Nisbett J.K. (2015). *Shigley's Mechanical Engineering Design*, 10th ed. New York: McGraw-Hill.

Cauchy A.-L. (1821). *Cours d'Analyse de l'École Royale Polytechnique, Première partie, Analyse Algébrique*. New York: Cambridge University Press.

Chandrasekaran R. and Tamir A. (1990). Algebraic optimization: The Fermat–Weber location problem. *Mathematical Programming*, 46(2), 219–224.

Childs P.R.N. (2014). *Mechanical Design Engineering Handbook*. Amsterdam: Elsevier.

Cloud M., Byron C. and Lebedev L.P. (1998). *Inequalities: With Applications to Engineering*. New York: Springer-Verlag.

Collins J.A. (2003). *Mechanical Design of Machine Elements and Machines*. New York: John Wiley & Sons.

Cover T. (1987). Pick the Largest Number, in *Open Problems in Communication and Computation*, Edited by T.M. Cover and B. Gopinath, New York: Springer-Verlag, 152–152.

Cvetkovski Z., (2012), *Inequalities Theorems, Techniques and Selected Problems*. Berlin: Springer-Verlag, Berlin.

DeGroot M.H. (1989). *Probability and Statistics*, 2nd ed. Reading, MA: Addison-Wesley Publishing Company.

DeVoe H. (2012). *Thermodynamics and Chemistry*, 2nd ed. Englewood Cliffs, NJ: Prentice-Hall.

Dhillon B.S. (2017). *Engineering Systems Reliability, Safety, and Maintenance*. Boca Raton, FL: CRC Press.

Dohmen K. (2006). Improved inclusion-exclusion identities and Bonferoni inequali-ties with reliability applications. *SIAM Journal on Discrete Mathematics*, 16(1), 156–171. DOI: 10.1137/S0895480101392630

Easley D. and Kleinberg J. (2010). *Networks, Crowds, and Markets: Reasoning about a Highly Connected World*. New York: Cambridge University Press.

Ebeling C.E. (1997). *Reliability and Maintainability Engineering*. Boston, MA: McGraw-Hill.

Engel A. (1998). *Problem-Solving Strategies*. New York: Springer.

Fink A.M. (2000). An essay on the history of inequalities. *Journal of Mathematical Analysis and Applications*, 249, 118–134.

Florin L., Petru C. and Costin B. (2020). Optimization methods for redundancy allocation in large systems. *Vietnam Journal of Computer Science*, 7(3), 1–19.

Floyd T.L. and Buchla D.L. (2014). *Electronic Fundamentals, Circuits, Devices and Applications*, 8th ed. London, UK: Pearson Education Limited.

French M. (1999). *Conceptual Design for Engineers*, 3rd ed. London: Springer-Verlag Ltd.

Fu H.M., Kang Y.M. and Shen H.G. (2010). Collection efficiency of fibrous filter on dust load. In 2010 4th International Conference on Bioinformatics and Biomedical Engineering (pp. 1–5).

Gere J.M. and Timoshenko S.P. (1999). *Mechanics of Materials*. Cheltenham, UK: Stanley Thornes Publishers.

Gullo L.G. and Dixon J. (2018). *Design for Safety*. Chichester, UK: John Wiley & Sons.

Hardy G., Littlewood J.E. and Pólya G. (1999). *Inequalities*. Cambridge Mathematical Library, Cambridge, UK: Cambridge University Press.

Hearn E.J. (1985). *Mechanics of Materials*, 2nd ed., vol. 1. Oxford: Butterworth-Heinemann.

Henley E.J. and Kumamoto H. (1981). *Reliability Engineering and Risk Assessment*. Englewood Cliffs, NJ: Prentice-Hall.

Hibbeler R. (2019). *Structural Analysis in SI Units, Global edition*, London: Pearson Education.

Hill S.D., Spall J.C., and Maranzano C.J. (2013). Inequality-based reliability estimates for complex systems. *Naval Research Logistics*, 60(5), 367–374.

Horowitz P. and Hill W. (2015). *The Art of Electronics*, 3rd ed. Cambridge, UK: Cambridge University Press.

Hoyland A. and Rausand M. (1994). *System Reliability Theory*. New York: John Wiley and Sons.

Kaplan S. and Garrick B.J. (1981). On the quantitative definition of risk. *Risk Analysis*, 1(1), 11–27.

Kazarinoff N.D. (1961). *Analytic Inequalities*. New York: Dover Publications.

Kiureghian A. (2022). *Structural and System Reliability*, Cambridge: Cambridge University Press.

Koziołek S., Chechurin L. and Collan M. ed. (2018). *Advances and Impacts of the Theory of Inventive Problem Solving: The TRIZ Methodology, Tools and Case Studies*. Berlin: Springer.

Kundu C. and Ghosh A. (2017). Inequalities involving expectations of selected functions in reliability theory to characterize distributions. *Communications in Statistics – Theory and Methods*, 46(17), 8468–8478.

Levi M. (2009). *The Mathematical Mechanic*. Princeton, NJ: Princeton University Press.

Lewis E.E. (1996). *Introduction to Reliability Engineering* 2nd ed., New York: John Wiley & Sons.

Livio M. (2009). *Is God a Mathematician?* New York: Simon and Shuster Paperbacks.

Long Chen, Yanlai Zhang, Zuyong Chen, Jun Xu and Jianghao Wu (2020). Topology optimization in lightweight design of a 3D-printed flapping-wing micro aerial vehicle. *Chinese Journal of Aeronautics*, 33(12), 3206–3219.

Maier A., Oehmen J. and Vermaas P.E. (2022). *Handbook of Engineering Systems Design*. Berlin: Springer.

Makri F.S. and Psillakis Z.M. (1996). Bounds for reliability of k-within two-dimen-sional consecutive-r-out-of-n failure systems. *Microelectronics Reliability*, 36(3), 341–345.

Mandolini M., Pradel P. and Cicconi P. (2022). *Design for Additive Manufacturing: Methods and Tools. Applied Science (MDPI)*, 12(13): 6548.

Mannaerts S.H. (2014). Extensive quantities in thermodynamics. *European Journal of Physics*, 35(2014), 1–10.

Marshall A.W., Olkin I. and Arnold B.C., (2010), *Inequalities: Theory of Majorization and its Applications*, 2nd ed. New York: Springer Science and Business Media.

Matthews C. (1998). *Case Studies in Engineering Design*. London: Arnold.

Miller I. and Miller M. (1999). *John E. Freund's Mathematical Statistics*, 6th ed. Upper Saddle River, NJ: Prentice Hall.

MIL-STD-1629A (1977). *US Department of Defence Procedure for Performing a Failure Mode and Effects Analysis*. Washington, DC: US Department of Defence.

Modarres M., Kaminskiy M.P. and Krivtsov V. (2017). *Reliability Engineering and Risk Analysis, A Practical Guide*, 3rd ed. Boca Raton, FL: CRC Press.

Mott R.L., Vavrek E.M. and Wang J. (2018). *Machine Elements in Mechanical Design*, 6th ed. Upper Saddle River, NJ: Pearson Education.

Newman M.E.J. (2007). Power laws, Pareto distributions and Zipf's law. *Contemporary Physics*, 46(5), 323–351.

Norton R.L. (2006). *Machine Design, An Integrated Approach*, 3rd ed. Upper Saddle River, NJ: Pearson International.

O'Connor P.D.T. (2002). *Practical Reliability Engineering*, 4th ed. New York: John Wiley & Sons.

Orloff M. (2006). *Inventive Thinking Through TRIZ*. 2nd ed. Berlin-Heidelberg: Springer.

Pachpatte B.G. (2005). *Mathematical Inequalities, North Holland Mathematical Library*, vol. 67, Amsterdam: Elsevier.

Pahl G., Beitz W., Feldhusen J. and Grote K.H. (2007). *Engineering Design*. Berlin, Germany: Springer.

Papadimitriou C.H. and Yannakakis M. (1982). *Combinatorial Optimization: Algorithms and Complexity*. Englewood Cliffs, NJ: Prentice-Hall.

Penrose R. (1989). *The Emperor's New Mind*. Oxford, UK: Oxford University Press.

Podder D. and Chatterjee S. (2022). *An Introduction to Structural Analysis*. Boca Raton, FL: CRC Press.

Pop O. (2009). About Bergström's inequality. *Journal of Mathematical Inequalities*, 3(2): 237–242.

Qiao H. and Li H. (2013). The discussion on optimization models of pure bending beam. *International Journal of Advanced Structural Engineering*, 5, 11.

Ramakumar R. (1993). *Engineering Reliability, Fundamentals and Applications*. Upper Saddle River, NJ: Prentice Hall.

Rantanen K. and Domb E. (2008). *Simplified TRIZ*, 2nd ed. New York: Auerbach Publications.

Rastegin A. (2012). Convexity inequalities for estimating generalized conditional entropies from below. *Kybernetika*, 48(2), 242–253.

Rosenbaum R.A. (1950). Sub-additive functions. *Duke Mathematical Journal*, 17, 227–247.

Rosenthal S., Maaß F., Kamaliev M., Hahn M., Gies S. and Tekkaya A.E. (2020). Lightweight in automotive components by forming technology. *Automotive Innovation* 3, 195–209.

Rozhdestvenskaya T.B. and Zhutovskii V.L. (1968). High-resistance standards. *Measurement Techniques*, 11, 308–313.

Samuel A. and Weir J. (1999). *Introduction to Engineering Design: Modelling, Synthesis and Problem Solving Strategies*. London, UK: Elsevier.

Sedrakyan H. and Sedrakyan N. (2010). *Algebraic Inequalities*. Cham, Switzerland: Springer.

Shubin Si, Jiangbin Zhao, Zhiqiang Cai and Hongyan Dui (2020). Recent advances in system reliability optimization driven by importance measures. *Frontiers of Engineering Management*, 7(3), 335–358.

Steele J.M. (2004). *The Cauchy-Schwarz Master Class: An Introduction to the Art of Mathematical Inequalities*. New York: Cambridge University Press.

Su Y. and Xiong B. (2016). *Methods and Techniques for Proving Inequalities*. Singapore: World Scientific Publishing.

Tegmark M. (2014). *Our Mathematical Universe*. London, UK: Penguin books.

Thompson G. (1999). *Improving Maintainability and Reliability through Design*. London, UK: Professional Engineering Publishing.

Tipler P.A. and Mosca G. (2008). *Physics for Scientists and Engineers: With Modern Physics*. New York: W.H. Freeman and Company.

Todinov M. (2023b). Enhancing the reliability of series-parallel systems with multiple redundancies by using system-reliability inequalities. *ASCE-ASME Journal of Risk and Uncertainty in Engineering Systems, Part B: Mechanical Engineering*, 9, 041202.

Todinov M. (2024a). Lightweight designs and improving the load-bearing capacity of structures by the method of aggregation. *Mathematics*, 12, 1522.

Todinov M.T. (2002a). Statistics of defects in one-dimensional components. *Computational Materials Science*, 24, 430–442.

Todinov M.T. (2002b). Distribution mixtures from sampling of inhomogeneous microstructures: Variance and probability bounds of the properties. *Nuclear Engineering and Design*, 214, 195–204.

Todinov M.T. (2003). Modelling consequences from failure and material properties by distribution mixtures. *Nuclear Engineering and Design*, 224, 233–244.

Todinov M.T. (2006a). Equations and a fast algorithm for determining the probability of failure initiated by flaws. *International Journal of Solids and Structures*, 43, 5182–5195.

Todinov M.T. (2006b). Reliability analysis of complex systems based on the losses from failures. *International Journal of Reliability, Quality and Safety Engineering*, 13(2), 1–22.

Todinov M.T. (2007). *Risk-Based Reliability Analysis and Generic Methods for Risk Reduction*. Amsterdam: Elsevier.

Todinov M.T. (2013). New models for optimal reduction of technical risks. *Engineering Optimization*, 45(6), 719–743.

Todinov M.T. (2016). *Reliability and Risk Models: Setting Reliability Requirements*, 2nd ed. Chichester, UK: John Wiley & Sons.

Todinov M.T. (2017). Reliability and risk controlled by the simultaneous presence of random events on a time interval. *ASCE-ASME Journal of Risk and Uncertainty in Engineering Systems, Part B: Mechanical Engineering*, 4(2), 021003. DOI: 10.1115/1.4037519

Todinov M.T. (2019a). Domain-independent approach to risk reduction. *Journal of Risk Research*, 23, 796–810.

Todinov M.T. (2019b). *Methods for Reliability Improvement and Risk Reduction.* Hoboken, NJ: John Wiley & Sons.

Todinov M.T. (2019c). Improving reliability and reducing risk by using inequalities. *Safety and Reliability*, 38(4), 222–245.

Todinov M.T. (2020a). *Risk and Uncertainty Reduction by Using Algebraic Inequalities.* Boca Raton, FL: CRC Press.

Todinov M.T. (2020b).Using algebraic inequalities to reduce uncertainty and risk. *ASCE-ASME Journal of Risk and Uncertainty in Engineering Systems, Part B: Mechanical Engineering*, 6(4). DOI: 10.1115/1.4048403

Todinov M.T. (2020c). Reducing uncertainty and obtaining superior performance by segmentation based on algebraic inequalities. *International Journal of Reliability and Safety*, 14(2/3), 103–115.

Todinov M.T. (2020d). Reliability improvement and risk reduction by inequalities and segmentation. *Proceedings of the Institution of Mechanical Engineers, Part O: Journal of Risk and Reliability*, 234(1), 63–73.

Todinov M.T. (2021a). On two fundamental approaches for reliability improvement and risk reduction by using algebraic inequalities. *Quality and Reliability Engineering International*, 37, 820–840.

Todinov M.T. (2021b). Generation of new knowledge and optimisation of systems and processes through meaningful interpretation of sub-additive functions. *International Journal of Mathematical Modelling and Numerical Optimisation*, 11(4), 428–448.

Todinov M.T. (2022a). A general class of algebraic inequalities for generating new knowledge and optimising the design of systems and processes. *Research in Engineering Design*, 33(2022), 161–171.

Todinov M.T. (2022b). Optimising processes and generating knowledge by interpreting a new algebraic inequality. *International Journal of Modelling Identification and Control*, 41(1–2), 98–109.

Todinov M.T. (2022c). Meaningful interpretation of algebraic inequalities to achieve uncertainty and risk reduction. *Proceedings of the Institution of Mechanical Engineers, Part O: Journal of Risk and Reliability*, 236(5), 841–854.

Todinov M.T. (2022d). Optimised design of systems and processes using algebraic ineualities. *Proceedings of the Institution of Mechanical Engineers, Part O: Journal of Risk and Reliability*, 236(8) 3912–3921.

Todinov M.T. (2023a). Reliability-related interpretations of algebraic inequalities. *IEEE Transactions on Reliability*, 72(4), 1515–1522.

Todinov M.T. (2023c). Can system reliability be predicted from average component reliabilities? *Safety and Reliability*, 42(4), 214–240.

Todinov M.T. (2023d). Probabilistic interpretation of algebraic inequalities related to reliability and risk. *Quality and Reliability Engineering International*, 39, 2330–2342.

Todinov M.T. (2024b). Reverse engineering of algebraic inequalities for system reliability predictions and enhancing processes in engineering. *IEEE Transactions on Reliability*, 73(2), 902–911.

Ugural A.C. (2022). *Mechanical Engineering Design (SI Edition).* Abingdon: CRC Press.

Valdes J.E., Zequeira R.I. (2006). On the optimal allocation of two active redundancies in a two-component series system. *Operations Research Letters*, 34, 49–52.

Vedral V. (2010). *Decoding Reality.* Oxford, UK: Oxford University Press.

Vose D. (2000). *Risk Analysis, A Quantitative Guide*, 2nd ed. New York: John Wiley & Sons.

Walton D. and Moztarzadeh H. 2017. Design and development of an additive manufactured component by topology optimization. *Procedia CIRP*, 60, 205–210.

Wang C. (2020). *Structural Reliability and Time-Dependent Reliability.* Cham, Switzerland: Springer Series in Reliability Engineering.

Weiszfeld E. (1937). Sur le point pour lequel la somme des distances de n points donnes est minimum. *The Tōhoku Mathematical Journal*, 43, 355–386.

Wesolowsky G. (1993). The Weber problem: History and perspective. *Location Science*, 1, 5–23.

Wigner E. (1960). The unreasonable effectiveness of mathematics in the natural sciences. *Communications in Pure and Applied Mathematics*, 13(1).

Winkler R.L. (1996). Uncertainty in probabilistic risk assessment. *Reliability Engineering and System Safety*, 85, 127–132.

Wolfson R. (2016). *Essential University Physics*, 3rd ed. Upper Saddle River, NJ: Pearson.

Xie M. and Lai C.D. (1998). On reliability bounds via conditional inequalities. *Journal of Applied Probability*, 35(1), 104–114.

Yang L., Harrysson O.L.A., Cormier D., West H., Zhang S., Gong H. and Stucker B. (2016). Design for Additively Manufactured Lightweight Structure: A Perspective, Solid Freeform Fabrication 2016: Proceedings of the 26th Annual International Solid Freeform Fabrication Symposium – An Additive Manufacturing Conference.

Yanga K-H and Ashourb A.F. (2011). Aggregate interlock in lightweight concrete continuous deep beams. *Engineering Structures*, 33, 136–145.

Yi Ding Yi, Yishuang Hu and Daqing Li (2021). Redundancy optimization for multi-performance multi-state series-parallel systems considering reliability requirements. *Reliability Engineering and System Safety*, 215(2021), 107873.

Yong Su, and Bin Xiong. (2016). *Methods and Techniques for Proving Inequalities*, Beijing, China: East China Normal University Press and World Scientific Publishing.

Index

Printed in the United States
by Baker & Taylor Publisher Services